The Mr. Gadget® Consumer Tech Guide

Volume One

Electronics, Gadgetry & Technology:

The One and Only Mr. Gadget

Reveals What to Buy and Why to Buy It!

STEVE KRUSCHEN

Library of Congress Cataloging-in-Publication Data

Library of Congress Catalog Number: 2012938475

Copyright © 2012 Steve Kruschen
ISBN-10: 0-9842085-5-0
ISBN-13: 978-0-9842085-5-5

Publisher: CFBP Bestsellers
An imprint of CFB Productions Inc.
P. O. Box 50008, Henderson, NV 89016
www.CFBPBestsellers.com

DEDICATION

My dear, sweet, sometimes even patient wife of more than 36 years, Jan, or as I often refer to her, *The Long Suffering Mrs. Gadget*, well deserves this dedication more than any other. She, along with my long suffering [now] adult children, has had to endure a life quite different and more chaotic than with most husbands and fathers.

This stuff is my passion, my work, and, I must say, even my hobby! It is what I enjoy; researching and having hands-on with myriad products and technologies. Let me just say it has not been easy on any of them. So, to Jan, Douglas, Marnie and Rhonda, I dedicate this, Volume One, my first book, with thanks and my undying love and appreciation for your support, *your* love and help at every step along the way.

I would not be half the man I am without you all,
and isn't *that* a scary thought!

CONTENTS

ACKNOWLEDGMENTS

I must first acknowledge my personal manager of 20 years, Clinton Ford Billups Jr., for it was he who encouraged, cajoled and prodded me, both gently and with more than a little pressure, to do this in the first place. How, I asked, can I write a book about technology and products that change all the time? By the time I write it, it's going to be obsolete. We figured it out, that it would be best to focus on the products and technologies that endure and do not change either at all or dramatically over time. With a healthy dose of hands-on experience, provide the advice asked for over many, many years, but with a nod to what is new, now.

Through modern technology and print-on-demand service, I am able to write this book and others to come, with revisions and additions, and publish as I see fit. Instead of a very long lead-time in traditional publishing, from the time this book was finished and formatted for printing, it is but a few short weeks before becoming available to readers. Revisions, even corrections, are easy to do on the fly as needed. You order the book, and only then is it printed and delivered, or it can be available in local bookstores that have ordered it to stock in their stores. Amazing!

So, thank you, Clinton, for your guidance, direction and support and all other manner of help getting to this point, this FIRST book finished . . . Finally!

My next acknowledgment, of course, must go to my wife and children for their love and support, including Jan's and Rhonda's editing skills. I love you all so very much and thank you for your immeasurable contributions.

i

CYA

INTRODUCTION

When I'm eating dinner in a restaurant, the most frequent question that I am asked is not, "May I take your order?" Instead other diners, bus or wait staff or even perfect strangers who have seen me through the restaurant window will approach me to ask, "Aren't you that gadget guy that I saw on television this morning?"

For more than three decades, I have made more than 1,000 appearances on national and local TV news shows to help viewers make intelligent decisions when it comes to purchasing consumer electronics, gadgets and new technology.

So, it's not surprising that the second most frequent question that I'm asked in restaurants is usually along the line of "What kind of cell phone do you think I should buy?" or "What's the best camera to buy?" or "Should I buy a desktop, laptop or a tablet computer?" or… Well, you get the point!

There's often no easy or quick answer to these questions since every consumer has different needs and expectations. However, in the hopes of answering some of these questions (and maybe eating a meal before it gets cold), I decided to share my recommendations in this book, which I hope you'll find helpful in making your buying decisions.

However, if you do see me in a restaurant, at an airport or just walking on the street, please still don't hesitate to say "Hi!" and ask your question, which I'll do my best to answer. My personal manager would prefer that you buy this book since I don't pay him a commission on free advice. Is this why he tells me that he doesn't like to go to out to dinner with me because of all these interruptions? No, I think he's just afraid that he will have to pick up the check!

Steve Kruschen
The One & Only Mr. Gadget®
Los Angeles, CA

CHAPTER 1

TELEVISION

PLASMA, LCD, 3D, PROGRAMMING CHOICES, SET-TOP BOXES AND MUCH MORE!

Once upon a time, buying and watching TV was simpler; choose the picture size and either color or black and white. Maybe there was a bit more to consider, but not much. It was a matter of price and extras – did the TV come with a radio and record player, for example? Or was it to be a console or portable?

Today, everything is wide screen and high definition. In addition to the basic technology decision between plasma and LCD, there is also the new 3D component to consider. And the issue of size on the new sets is one of THE most confusing considerations of them all.

First, however, is the need to know how it is the user is to receive the signal.

There are four ways a signal can reach the TV.

The **most common *old-school* method** is traditional *over-the-air* reception using an antenna. Yes, in many areas of the country, users may be able to receive *broadcast* stations in this oldest of methods in which the stations send

1

signals through the air. A rooftop or inside antenna receives the signal. The antenna's wire, a piece of round coaxial cable, usually the same as or similar to the wire used if the consumer receives signal from a cable company, connects directly to the TV using its built-in digital tuner. It should be noted that all broadcast TV signals are digital.

When the changeover from analog to digital occurred and all over-the-air analog broadcasting ceased, everyone's old-style TV's internal analog tuner became useless. Users with those old TVs who received TV stations through an antenna had to purchase a digital converter box (or receive one free with a special coupon that was available at the time, but no longer), which takes the new digital broadcast signal and converts it to the old analog signal, then on to the TV. This method limits programming availability to just a few channels at most. These few channels are networks – ABC, CBS, Fox, NBC and, perhaps, a local PBS station – in addition to, possibly, a few other local stations. That's it.

Reception via a cable TV operator is the next option. In this scenario, the TV's built-in tuner is bypassed. Signal from a cable company is most often fed through the back of a cable box, also called a set-top box. A special cable called HDMI carries audio and video in its single connection from the box to an HDMI port on all high def TVs. If you still have an old, picture tube-type TV there are other connection capabilities, but here I will concentrate on the newer technologies. Channel selection is usually done by a remote control that changes channels in the cable box. High definition local channels may be included in the basic cable fees. The set-top box may also be capable of recording programming for playback at the user's choice of time and day.

Now is a good time to mention that **the first and still best such recorder, or DVR, is TiVo.** Did you know you likely have a choice and could purchase and use your own TiVo DVR? Don't expect your cable TV supplier to volunteer this information to you! They want you to pay a monthly fee for their non-TiVo, inferior DVR.

TiVo is powerful and enabling, with the best user interface, better than all other DVRs. Most users would rather fight than switch having experienced TiVo and then sampling other devices. TiVo boxes are premium products, not for their initial cost, starting at $100, so much as for their operating cost. Compatible with most cable systems and replacing the cable company's own set-top box, TiVo has its own *better* program guide with fees above and beyond what users pay for basic cable. In many cases, however, that cost is not much more than and may be less than what cable companies charge for their own DVR rental fee per month. TiVo DVRs can record a minimum of two programs simultaneously, while playing a pre-recorded program. The newest product, TiVo Premiere Elite, at $500 (all prices plus service), can record *four* shows at once and up to 300 hours of HD recording. Ah, there is much to love with a TiVo!

TiVo is much more than just a DVR, too. Added to the TV-watching and easy-to-set recording experience, TiVo is a *smart* DVR. Add your accounts with Pandora, Rhapsody, Netflix streaming, Blockbuster on Demand, Amazon Instant Video, Hulu Plus, and more, and search for and watch YouTube videos. Add the ability to stream music from a networked Mac or Windows PC and access to Web-based photo services, plus online and mobile scheduling of recordings, and still more. In homes with multiple TiVo units, users can transfer recording between networked TiVo units, so what was recorded in, say, the living room, may be watched through any other TiVo in the home. *TiVo is so much more than just a DVR!*

I must also remind everyone that any Internet-based streaming content through ANY device necessitates a speedy and reliable Internet connection of at least 3Mbps download capability, and preferably higher. Check your speed at speedtest.net.

Let me just say that I *love* and recommend TiVo above any other set-top box connected to a cable company. Users sacrifice video on demand (VOD) capabilities when using TiVo, however. This is a trade-off I gladly accept. My recommendation is to visit the TiVo Website at **TiVo.com**, look at their products, price them singly and with a multi-unit service discount, even with their lifetime service option, currently at $499, and compare to the **overall costs** of using cable providers' hardware, *even if a cable supplier's offering includes their own TiVo-partnered DVR*. Cable companies' TiVo-partnered DVRs do not allow all the wonderful added capabilities of stand-alone TiVo DVRs. I know this may be a bit confusing just reading this paragraph without the context of owning a TiVo DVR, but a visit to the TiVo Website and the cost comparisons should clear things up for you. **In addition to all that the TiVo site has to offer, also check with my friends at weaknees.com, the best TiVo repair and upgrade shop around**, serving users across the US. Weaknees also offers great advice and a range of upgrade options. I'm a very happy Weaknees customer!

Not everyone wishes to buy a TiVo DVR, which I understand, and not every cable TV system is compatible with TiVo, also understood.

Consumers, do what you must and what you like, of course. All I suggest is that the availability of TiVo for your situation be explored, and, if available, give it the arithmetic test to see how buying and using TiVo compares with what is currently in use. You just might be pleasantly surprised!

My current choice for TV service provider is Verizon FiOS, a cable TV competitor in many markets. Not everyone has this choice, as with TiVo, but here in SoCal where I live it has proven the best choice when considering value and performance. And I get to use my TiVo DVRs! If Verizon FiOS is in your market, please look into this option and give it the arithmetic test. I have never been so pleased with any TV and Internet service. FiOS service is the most reliable I've ever encountered. Polling colleagues with the service revealed similar impressions, unlike any other TV or Internet provider. The service is rarely down and the TV pictures are better than from any other cable supplier and better than any satellite-based service. The reason? BANDWIDTH. It is in abundance with FiOS' fiber optic-based system, with more bandwidth than any competitive service. It's that simple.

Third, the TV signal might come from a satellite-based provider – DirecTV or Dish Network. Signal comes in from a satellite dish on the outside of the home or apartment and is sent to a set-top box similar in form and function to a cable TV box. TiVo used to partner with DirecTV, but no more. The company has said it would be back with TiVo, but that promise has slipped many times and years since the initial announcement. Users considering satellite TV are advised to simply look at cost and programming differences and make the choice based upon these factors. Both do a fine job. There are better sports packages with DirecTV and better international programming options with Dish. Pick your poison!

Let me sidestep yet again to recommend an excellent *gadget* for many TV viewers. Do you miss out on favorite sports events or virtually ANY TV programs while away from home? Well, do you? If you would like to watch YOUR home TV while anywhere away from home, this is for you! **Slingbox is the gadget,** with info at **Slingbox.com**. Starting at about $180, watch and *control* YOUR live TV at home, i.e., change channels, or control and watch your at-home DVR programs on a desktop, laptop, iPad or smartphone anywhere you happen to be. It's quite simple. Slingbox is so much fun! Only Slingbox allows these capabilities, is easy to set up and easy to use. You don't have to be a geek to use Slingbox! Shop online for Slingbox.

4

Slingbox has competition from Belkin. If you like the idea of the simplest Slingbox devices, their Slingbox Solo (info at their Site), do check out Belkin's $150 **ScreenCast TV (belkin.com)**, which I viewed at the 2012 Consumer Electronics Show in Las Vegas in January. It appears to have similar, if not the same, capabilities for less money. Though not available at this writing, it may be by mid-2012.

Fourth, and growing in popularity, users may opt to see programming that is mostly *not live* (as it could be from the other three methods) but *streamed* over the Internet and seen on a TV. Consumers connect to the Internet via computer or another type of set-top device such as **AppleTV (apple.com/appletv), Boxee Box (boxee.tv), Logitech Revue with Google TV (logitech.com/en-us/smarttv)** and **Roku (roku.com)**, then, through the box, log onto already subscribed to services such as Hulu and Hulu Plus, Netflix, the iTunes Store in the case of AppleTV, and others to find desirable programming that is either free or fee-based. Programming can be watched on a computer screen without a box *or* an alternate device, such as these boxes, may be connected to a high def or standard definition (old school) TV.

Some content comes with a fee or subscription charge, such as movie rentals and some live sporting events where available and on specific devices, as well as some streaming music services. Weigh monthly cable or satellite charges against what might be spent in this alternative way. Many of you will be pleasantly surprised.

Among the four choices indicated immediately above, **Roku**, with three current models of Roku 2 boxes priced at $60, $80 and $100, is the most robust performer and most mature platform of its type. **Roku rocks!** And it is so inexpensive. The little products are not only inexpensive, but they are small in size, energy misers and have a growing roster of content partners. Investigate the products and content partners to see all the programming choices at **roku.com**. Prepare to be amazed!

Because of such a vast partner list and so many content options, I've read that many Roku users have cancelled their cable or satellite TV service and see what they want through their Roku box, supplemented with content found online through a computer. Remember that there is no way *YET* to watch LIVE shows on this or any of the other alternate choices. However, live TV news and previous news reports are available in most markets, found by searching online and viewing on the computer screen. Admittedly, this is not nearly as convenient as simply firing up the TV, choosing a channel through a set-top box and watching. But, saving from about $30 to well over $100 per month can be a comforting alternative. The combination of watching Roku partner content on a TV screen and other content on the computer screen leaves little not somehow available as if connected to cable or satellite.

Roku 2 XS, $100, is my recommendation. This top-of-the-line little box is distinct among the other Roku products in that it is the only one that can be connected via Ethernet as well as Wi-Fi. It also comes with a better remote control, perfect for games, including Angry Birds, which comes with the device. Use it with any other Roku-available game, too. In addition, this model has a USB port that works with USB sticks and USB hard drives to access some of the more common audio and video formats users may already have. More info is at the Roku Site. Finally, it shares with the $80 model, Roku 2 XD, the ability to play the best quality video, 1080p, as well as 720p. The $80 box is only recommended for users without Ethernet at the set, but keep reading for a solution to this conundrum. Spend the additional $20 for the best performance!

Roku is not the only player in the area of alternate set-top boxes with apps. Router maker Netgear (**netgear.com**) has a product series called **NeoTV**, which has apps and can connect and catalog users' media on their home network for "easy" playback on their nice TV. Hard drive makers Western Digital and Seagate have similar products. **WD TV** from Western Digital (**wdc.com**) and **GoFlex TV** from Seagate (**seagate.com**) are both media players in this space. Check them out to determine if any meet your interests and needs. These differ from Roku in their design to be aggregators of content on your home network. They also are designed to have attached local storage (a portable hard drive, for instance), on which may reside photos, music, and home videos from camcorders or even old VHS tapes converted for viewing in this arena. Difficulty may come in the form of incompatible "codecs," the bit of software that looks at the video (with its audio component) file data and makes it play through the attached device. I have considerable experience with this.

I have years of material, from old home video to new, and from DVDs long ago "ripped" (copied) to my hard drive. Over the years, the software that converts, for example, DVDs to useable data that is no longer on a DVD so I can play the movie from a hard drive, has changed. The file extensions, perhaps mv4, m4v or even mp4, have given fits to the devices I've played with in this category. There is not one standard format under those extensions, unfortunately. I thought an mv4 and the others, plus still others would be the same wherever I saw the extension, but no! And some raw footage taken directly from some camcorders cannot be played through these devices. I'll see an error message indicating there is no capability to play the audio, for example. That is the problem with some of these named devices. Makers are working hard toward increased compatibility with formats, but it is almost a moving target. Western Digital representatives have told me the company's WD TV product has the greatest capability for success with more codecs than competitors.

I could always play my copied videos and camcorder files on a computer, but not necessarily through one of these devices and it has been frustrating.

However, if what I have just learned and seen at the 2012 Consumer Electronics Show in Las Vegas is as I expect, all of this incompatibility may be coming to an end!

The message is simple – If you can see it on your computer, first on a Windows PC, then the image or video can be seen, *wirelessly*, on a nearby TV. This feat is courtesy of **Intel** technology first shown in 2010 called **WiDi** (**Wireless Display**) and built into many new Windows laptops, into accessory, standalone products, in TVs and in yet-to-be other products.

Consumers with a WiDi-equipped PC, but with the capability NOT built in on the TV side will need an external TV adapter. Get one from **Belkin** (**ScreenCast TV** adapter – **bit.ly/xu4wxu** with link shortened for your convenience), **D-Link** (**DHD-131** TV adapter – **dlink.com/mainstage**), and **NETGEAR** (**netgear.com/ptv**). At the very least, **Belkin** will have a two-piece solution in their ScreenCast TV line with an adapter for the PC and one for the TV.

Apple will surely have its own implementation of these capabilities. Characteristically, the company has not made an announcement on the subject.

The ability to get video (and audio and photo) content *ultimately easily* streamed wirelessly from a computer and onto a TV should make it much easier to enjoy individual users' or family content through the big screen. It is my understanding that translations and incompatibilities are, therefore, avoided. This should be exciting news to all consumers who would like to hear and see their *stuff* with the easiest to use, simplest process from their in-home computer network on their nice, big TV screen.

Also at CES, **Technicolor** announced their FREE **M-Go** app software solution (**www.m-go.com**), compatible with WiDi and platform independent. It seems to me that this may be among the best ways to catalog media and then "throw it to the TV screen of choice" in one's home. Software is for computers as well as mobile devices. Imagine sitting on the couch, smartphone or tablet in hand, watching TV. Then, picking up said mobile device, launching the M-Go app and touch-navigating through the possibilities on your home network AND outside to wherever the source of your interest is available. Ponder this last point a moment.

What if you wanted to see True Grit from 1969 (the example my M-Go guide used at the CES demo)? M-Go finds it anywhere it exists, on your network or someplace on the wild and wooly Internet, perhaps the DVD for purchase or the video may be available to watch via streaming capability from Netflix. As I have a Netflix account and would have already tied my M-Go to my Netflix account (or it could be Amazon Instant Video, Hulu Plus or another outlet or service), I could choose to purchase the item then and there OR touch to stream the content. Then, it is a matter of choosing the screen of choice to throw the image onto, so to speak. Technicolor is not totally altruistic. They will also show what you are asking for from their own catalog of content. They just want the first crack at your business if they have what you want, and you are still not obligated to buy from them. M-Go has me excited and, I hope, you as well.

Boxee Box is best suited for geeks and is worth a look online, allowing potential users to see and learn about it. **AppleTV**, though easy to set up, does not have quite the number of content of partners as does Roku, but it is still worth a look, for both Windows computers and Macs.

This may change in 2012 or 2013 if speculation of a true Apple-branded TV should become reality. Maybe it will become possible to receive LIVE and pre-recorded content through Apple, on their TV and on numerous other devices, *bypassing* cable and satellite providers. It can all come via Internet, so long as users have robust enough speed and reliability to allow for such streaming. Of course, it would come with fees that would be less than current providers charge. We can only hope! Technologically, other than the issue of Internet speed and bandwidth in general, there are no obstacles to receiving TV programming without cable or satellite companies dictating terms that are as expensive as they are today. I would love to give it a try!

Logitech Revue with Google TV has promise and is fun to use, different from all the others, though not without problems. The product differs from competitors, as it is an intermediary device between a cable box, TiVo or satellite TV box and the TV. Signal passes from cable or satellite through the Logitech Revue box and on to the TV. In this way, it works *with* cable or satellite service and not as a replacement for either. I like it! I like that it comes with a wireless keyboard that acts as a remote control for TV and cable or satellite boxes. It runs on Google's Android operating system and includes its own Google Chrome browser, allowing users to visit Websites directly on the TV through the Google TV box. It is not a competitor to the other, recommended devices as it is not a way to eliminate the need for and cost of a cable box or satellite set-top box. It features many useful apps. Be sure to check it out, however, at the Website indicated above.

Here is the biggest downside to *this* device. While it is the ONLY simple way to see the Internet on your TV as you would like to do if you connected and used your computer through a TV, it does NOT allow users to watch shows on network Websites. A somewhat anemic Intel Atom processor of the type used in first-generation netbooks powers Logitech's Google box. It can keep up with *most* tasks.

Some slowdown is inevitable when surfing and streaming videos from this device on the TV. Go to, say, cbs.com and see the links to watching shows online. It's a wonderful capability, also offered by the other networks. Unfortunately, and for reasons beyond me, this is prohibited through the Google TV device! In my view, this is a deal breaker. It is NOT Logitech's or Google's fault. It is the networks' insistence on not allowing these sites to stream through this device. It is a deal breaker because absent this capability, it may not be worth the price of admission, though only $100.

Though *this* Logitech device and its cousin from Sony are (or, rather, were) the initial product salvo for the Google TV concept, 2012 will see renewed interest on the part of consumer electronics partners, putting Google TV inside TVs and in new and more exciting, higher performance standalone boxes. Performance will no longer be sluggish and there will be more content partners. *However,* unless and until such a product is permitted to stream either or both live TV without requiring cable *and* previously broadcast shows from the network Sites, Google TV will still have its deal-breaker incapability for most consumers.

Web browsing is not a Roku feature, but Roku boxes offer so many other benefits and features that it just is not practical for *most* users interested in such products to have BOTH Roku and Logitech Revue with Google TV. If you like what you see, then far be it from me to say not to get it. For the record, I like it and use Logitech Revue *in addition to* a Roku box. Finally, don't write it off. Check it out and see if and when it offers apps through the Android Market that may represent or increase its value and benefit to YOU, making it a worthwhile purchase.

Moving on, a computer's DVI or HDMI interface can connect via a cable to an HDMI port on the TV. Coming from a DVI-equipped computer, a DVI to HDMI adapter is needed. In the case of computer HDMI and DVI interfaces, sound is NOT carried over these connections in all computers, so a separate wire from the computer's audio out jack to the TV is needed. Confused? This can become more complicated and is more a geek's domain than for the non-technical person. If you are the latter, you may need to elicit the aid of a geek to try and make this work. The meaning of all this is that the TV becomes, in this instance, an external monitor for the computer.

On the other hand, it may be possible to simply connect a laptop or desktop computer directly to a TV's HDMI input *WITH* sound, to get all the Internet-sourced content, including streaming from Hulu and all the networks. It all depends upon the computer being used. If this is an attractive option, check with the computer maker to see if it can work for your situation. I know, for example, that certain Macs can do this with a cable adapter from the mini display port into a readily available adapter made by Moshi, and then through an HDMI cable on to the TV. Info is at **bit.ly/vCUEV7 (link shortened for your convenience)**. NOTE: I have NOT tried this device. I have read numerous reviews expressing concerns about the quality of this product. *Caveat emptor!*

As all "boxes" connect with the TV via HDMI, as do game devices and Blu-ray DVD players, it is also important to tell you how to get the best value when purchasing HDMI cables. **DO NOT BUY HDMI CABLES FROM** *MOST* **RETAIL STORES. You WILL pay way too much.** Head over to **monoprice.com** to find their **standard HDMI cables in a link on the left of the home page. Click it**! Find the shortest length you can get away with and note prices from under $2 to a paltry $5 for cables from 1.5 feet to 10 feet! Simply stated, digital info carried in an HDMI cable either flows or it does not. There is no such thing as a "better" cable in these lengths. **Don't waste money on expensive HDMI cables**! At **monoprice.com** low prices, why not get cables for ALL your HDMI-equipped products?

The only "better" cable might be one that is thinner, with a shorter connector and with greater flexibility. One or more of these might be a good idea in installations in which the TV is mounted flush against a wall, for example, where there is little clearance. Conventional HDMI cables are thick and not very flexible at the connector ends. Made by **Sanus (bit.ly/tpmnjd – link shortened for your convenience**), here is one example of thinner cable for those in need of this capability. Look online for **Sanus Super Slim HDMI Cables** to find availability and pricing. These are aesthetically better, **not** digitally better! Oh, and they are considerably more expensive, to the tune of about $35 for a six-footer.

A growing trend from TV makers is to have built-in Internet connectivity and content partners bring applications and programming directly to the TV without a separate box. I like the idea of that separate box, however, to provide me with the flexibility of programming sources the outside devices deliver, which is greater than on any current TV alone. This may change with successive generations of TVs, but if you have a TV now, an external source box, such as Roku, is the best way to go.

As stated above, with this method you cannot easily see TV shows and movies on TV as if viewing a live source. Think of it as if the programming was stored on a DVR and played back at the user's leisure. All this programming is digital, but not all of it is going to be in glorious high definition.

There are more and different ways for a computer to act as a DVR and to have a broadcast TV tuner or cable tuner connected from the outside, but these are more complicated and therefore not generally aimed toward the average consumer. Besides, full use of a computer connected to a TV, supplying programming and serving as a DVR would require that the computer be dedicated to this scenario and not used for other purposes. This is an expensive option!

For those who want the tuner-at-computer option, Windows and Mac users should go to **elgato.com** and see the **EyeTV Hybrid**. It is among the simplest solutions available. Note, however, that since a coaxial cable connects directly from the wall to this device, it can only receive cable TV channels that are available without encryption. It is NOT a cable box! This means it will likely only receive network feeds and maybe a few cable-only stations. Forget about receiving HBO and the rest of the higher-tiered channels, as these are most assuredly encrypted, requiring a cable company (or TiVo) box, at added expense, of course. I'm trying to suggest money-saving alternatives, and this one is not going to save money, unless you only want to see and record unencrypted TV signals on your computer. It is also becoming more common for cable suppliers to encrypt even local channels, so users may not get what is anticipated or expected using one of these. In other words, be sure the product is returnable if it proves to not work for you.

An additional source of programming might come from a Blu-ray DVD player connected via HDMI to the TV. Modern TVs have at least two HDMI ports, and as many as four in some of the larger models. In this way, users may watch standard and high definition Blu-ray DVDs on the "nice" TV as well as other sources of programming connected through other HDMI ports *and* directly from the Internet using built-in capabilities.

Could you be among those considering eliminating the current monthly cable or satellite fee?

If the principle TV viewers are not blown away by superb images (as can be seen on demo sets in stores or at others' homes), it would be a waste of money to go out and get an expensive high definition set. It just makes no sense. Also, for only occasional viewing the user may wish to shop by price and not by size and performance.

Alternatively, if viewers are dumbstruck by the clarity and sharpness, plus detail of a great picture, then, by all means, and if budgets allow, go for the good stuff. But **what *is* the good stuff?**

Plasma is best. Period. Sure, most of the sets available today are based on LCD technology. An increasing number of these are backlit using LED technology instead of the older, less expensive fluorescent technology, essentially a fluorescent light bulb behind the screen. **There is no such thing as an "LED TV," though ads would have you thinking differently.** *But, why, Mr. Gadget®?*

It's all marketing hype, I am afraid! Plasma technology is capable of delivering more and better-looking colors and contrast levels than LCD technology. LCD technology may require less energy, though even this is changing with modern plasma sets. Many are at parity with similar-size LCD TV power consumption, or are so close as to suggest this is no longer an issue.

Under some circumstances LCD TVs can deliver a more *pleasing* picture. This unusual circumstance occurs when there is sunlight streaming directly onto the TV screen. Absent that, and especially in rooms that I think of as having "normal lighting" without the extremes of direct sunlight on the screen, the more pleasing picture from plasma will almost always be chosen by consumers over images on an LCD TV. Do not mistake apparent brightness for sharpness, clarity and a brilliant color palette. Most TVs on retailers' shelves are set at levels too bright for home use. This is because our brains perceive that such brightness is a desirable characteristic. I can assure you, however, that once you get that new TV set up at home and adjust the various levels, you will discover that more can be seen, more subtlety of shading, for example, with the brightness set to lower levels.

One solution to achieving perfect display adjustment is to purchase a display with THX certification. Go to **thx.com**, and then drill through the links at the right under **FIND PRODUCTS**. Select **LG** and then **Panasonic** as **Manufacturers**, then **Displays** as **Product Type**, and then select either **THX** or **THX 3D Certified** under **THX Categories**. With one click on a THX setting on any THX-equipped TV, optimal video settings are dialed in automatically in a can't miss fashion. Note that only plasma sets are currently THX Certified in the US, though some high-end THX-certified LCD displays are on the way to our shores.

Regardless of the brand or price of your HD TV another way to achieve optimal picture quality is to use the calibration tool that is included on every THX-certified DVD movie disc. Look on the label for the THX logo! This built-in calibration guide is all that is needed to tweak a display to its optimal video settings, and it works well on all sets.

One easily discernable downside of LCD technology is the narrower "sweet spot" for best viewing. Plasma sets look good from a very wide angle to either side as well as high and low. Try this with LCD sets to see the significantly narrower *pleasing* viewing angles. Though manufacturers are working to mitigate this effect, plasma sets are still king in the viewing angle department.

Comparable size plasma and LCD set prices may also surprise you. Plasma costs have come way down, even though the technology is in far fewer sets. Depending upon the size, don't be surprised if the best plasma set you choose over LCD may also be the least expensive.

Some readers may recall long-ago concerns over plasma sets' long-term durability and the possibility of image burn-in. If you had not heard of either, and even if you had, simply don't be concerned. Today's plasma TVs are rated for a life of greater than 60,000 hours – that's more than 20 years of eight hours per day viewing, every day, all year. The burn-in issue is a thing of the past. As with any TV, however, leaving a static image on-screen for days on end is not a good idea.

Now, let's consider size. Plasma starts at 42-inches. Smaller than that, the choice must be LCD. Image quality in smaller sizes is not such a factor as in larger sets. Also, there are two current high definition standards to consider – 720p and 1080p. The higher numbers are better. Inasmuch as the brain cannot well discern 1080 quality in sets below about 40 inches, be guided accordingly. Even in larger screen sizes, if the viewer is at a distance exceeding, say, eight feet, it does not matter, so a 720p set will look just as good and cost less than a 1080p set! Note, too, that many broadcasters deliver only 720p content, including Fox TV and their sports coverage. Baseball and NASCAR races sure look great! NO content is broadcast in 1080p, only 1080i. Blu-ray DVDs and game consoles are where one would find 1080p content.

As a size guide, the best size for a distance of up to about seven feet from eyes to set is about 50-inches. Yes, this is a shocker, I know, but this is the way things work in the era of high definition. Farther away suggests a 60-inch or larger set. Larger than about 60-inches will likely mandate LCD, too, but they become quite expensive in larger sizes. If you have this specialty need for extreme picture dimensions, including dedicated home theater, seek source information from **hometheater.com** and **hdguru.com**. Both provide excellent, reliable content. The services of a custom installer may be in your future!

If you are not going to watch broadcast or cable stations in high definition, then I would encourage getting a smaller size as there will be less available picture information, less need for sharpness and contrast of better displays, as well as overall detail. Watching digital non-high definition on a large screen digital TV is, in my view, not so pleasing. I would not spend the money on a large screen in this usage model. Most TVs also have the ability to scale up from, say 480p (standard definition digital content) to 1080i or p. These TVs can also handle making standard DVDs look better than normal, closer to the higher standard than their native lower standard. Look into this feature on a new TV to be considered for purchase.

Now, which technology to choose? We're almost there, but first . . .

What about 3D? The sad thing is that the jury is still out on a single, winning 3D technology, even as manufacturers are clamoring to include it in their sets. **Passive 3D technology uses non-electronic glasses, like the ones experienced in movie theaters. Active technology requires electronic, rechargeable battery-operated glasses. There are at least two active technologies used today, so the glasses from one set maker may not be compatible with another set maker's active 3D technology. This will ease somewhat in 2012 when there will be a standard for active technology 3D glasses.** This is good! Did you know that such glasses are at additional cost after the purchase of this kind of 3D TV? It's like selling a car and only afterwards discovering that an additional purchase of tires is needed so it can actually be driven. Whose idea was this! Active glasses are coming down in cost, which is also a good thing. Some are as inexpensive as $30 each, so do your homework as to the added cost of ANY 3D set in order to actually participate in 3D programming.

Still, it is up to all manufacturers to adhere to this standard and, so far, not everyone is on board.

If, say, in eight years a consumer may wish to get additional or replacement glasses for a 2011 and earlier model 3D TV, glasses might not be available from the TV manufacturer and maybe not from outside companies. This is food for thought, at the very least.

And did you know that 3D does not work if you are in the habit of watching TV while lying down? The viewer *and* glasses must be in the same horizontal plane relative to the screen.

And what about content? Not much 3D programming is available today, though this is certain to change if 3D is ever to take off. There is the promise of much more native 3D TV programming in 2012. Converting non-3D, known as 2D programming, to 3D is not satisfying. 3D programming has to be created from the get-go as 3D for the experience to be worthwhile, at least with currently shown technology. *Currently*, DirecTV has the most dedicated 3D programming. You know this is going to change.

More thinking points about current (and maybe future) 3D TV efforts include the realization that a small percentage among the population simply cannot perceive 3D as intended. No, they are not ill and in daily living they are otherwise unaffected. Additionally, some individuals have reported feeling headachy, dizzy, and even sick to the stomach when watching 3D programming, though this is rare. Did you know that watching 3D programming is also best viewed and appreciated when the TV is viewed closer up than for 2D viewing?

Let's review – 3D costs more, provides *possibly* inconvenient viewing parameters and other possible downsides.

Do I recommend against the purchase of a set equipped with 3D technology? Friend and colleague, **HD Guru *Gary Merson* (hdguru.com)** put it in perspective for me. Said he, "3D TV is a *feature*. Consumers are free to use or not use this or any other *feature* on a TV. It is clear that more 3D TV models will appear in 2012," but there is no way to know if the feature will catch on to the extent that it will be around in five years and beyond on new TVs. **My recommendation is to not fight it, as 3D is probably going to be a feature on most available TVs.** Purchase glasses if and when you or others in your household want to use the *feature*.

My recommendation for plasma TV is to stick with the three major players, LG, Panasonic and Samsung. Each company, as Gary educated me (*AGAIN*), has an interest in providing great TV products that will bring you back to the brand again and again for the other products they make in addition to their TVs. These companies produce some or all of these – cameras, batteries, kitchen appliances, computers, air conditioners, Blu-ray players and many other products. If such a multi-tiered company *burns* a consumer who has bought anything significant, he or she is less likely to think favorably of that brand for ANY future purchases. On the other hand, if a TV-only or TV-primarily company lets down a consumer, there is a deep well of more consumers who will or might try the company for a TV.

In LCD TVs, I'd go with these same brands, plus having a look at Vizio, a top-selling brand. However, just because a brand sells well does NOT mean it is the best product nor will it last as long as other brands, such as those named above. Whoever it was that coined the phrase, "You get what you pay for!" may have been onto something.

Alternatively, I have heard first-hand from Vizio customers offering glowing praise for the company's customer service. When there have been issues under warranty, Vizio has stepped up to the plate to repair and, if needed, replace the set, simple as that.

What about after-warranty service and repair? This informative article from **HD Guru** is a **MUST read for anyone shopping for a new TV:** bit.ly/ug2TEi (link shortened for your convenience).

Hmmm, let's see: There is an HD TV, a set-top box from cable or satellite or TiVo, a Blu-ray player, perhaps a console gaming system, maybe a Roku and/or other box. **These set-top boxes all have one commonality in addition to being connected to the TV,** either directly or through an A/V receiver for surround sound. What is it? **All need Internet connectivity to function best!**

If you are just now putting a system together or even if you already have some of it, do you have the Internet at the area where the TV lives? If not, it becomes difficult to perform updates to the computer that really is your TV. What about "apps" that permit direct access to such things as Netflix, YouTube, Skype (with optional camera), LIVE weather and so much more. Blu-ray players are most easily updated with direct Internet connection, in addition to taking advantage of additional content available through many Blu-ray discs. Programming and software updates to a set-top box are faster than dial-up over a phone line, if that option is even available. Gaming consoles need the Internet. Roku and the other content-rich solutions in that area of technology rely on Internet connectivity to operate in any fashion.

It becomes a foregone conclusion that Internet needs to be where the equipment lives. Even if Wi-Fi is an option for ALL your "pieces," it is not the best way to bring Internet to the area. Wireless bandwidth is limited in any home and may suffer from too much by too many users at one time. Some are fortunate to have wired an Ethernet line to the area in question into a multiport switch, such as the **TrendNet** (trendnet.com) **TEG-S8** eight-port device I use. *Wired* needs do not impact *wireless* usage. Use wireless, if at all possible, only for mobile devices, such as phones, laptops and tablets and such, and for wireless-capable printers. Use wired connectivity for all other stationary devices whenever possible.

All Ethernet switches are simple, "dumb" boxes that allow one Ethernet wire's signal to be split and available to several devices, up to the number of ports available on the switch. Think of it as similar to using a multi-port USB hub at your computer to provide more USB connections, though it is not technologically the same (DON'T write to me about this!). The other end of the Ethernet connects to any available port on the user's router OR to another switch similarly connected at the router. TrendNet makes larger switches and a lesser capacity five-port model that is only about $6 less than the eight-port model that can easily be found for **$35** with **free shipping** at **amazon.com**. Get the eight-port switch or a larger one to accommodate expansion, making it future-proof.

This usage model is also another reason that the router serving up Internet should be of the gigabit Ethernet type and not the lower performance 10/100 speed. Gigabit Ethernet is up to 10 times faster. This is all in an effort toward being more future-proof! The reason for all that speed is to accommodate multiple streams of content to more than one device, perhaps several, at the same time. Be sure your Internet speed, in general, is up to the task! As suggested above, test it at **speedtest.net**. If you wish, ask your Internet supplier about availability and cost to provide faster service. Remember, that one Internet connection is shared with every one and every thing using it in your home. Unfortunately, not everyone has the availability of true high speed Internet. As well, not everyone wants or needs higher speeds. If you are happy with what you have and do not have to wait what, to you, is an unreasonable amount of time for things to load and complete, look no further! But things are changing, perhaps even for you, as you will read.

Internet switches require NO configuration. Just plug it or them in (they use very little power) and connect the Ethernet from the router, and then add inexpensive Ethernet cables from the switch to all the Ethernet connections on all your TV-related equipment at that one location. It IS that simple! **There is nothing to configure.** Ethernet cables are also inexpensively available from **monoprice.com**. Prices start at under $1 for up to two-foot cables available in several colors and just over $2 for 10-foot cables. Get **CAT6**, not CAT5, to further future-proof the installation, only because CAT6 is just about or exactly the same price as the older standard, CAT5. Most users will not see a difference between the two.

This is a good time to inspect your router and to look it up online. If yours is old and your or visitors' computers are not, it's probably time for an upgrade. This is especially so if your sphere of devices includes handheld or laptop devices using Wi-Fi, a high def TV, gaming devices, set-top boxes with Internet connectivity, Internet-based phones and who knows what else.

This is also a time to state the obvious, that the need for high speed Internet access is growing. If you are an occasional computer user, with older equipment and no younger and hipper folks at home, you might be fine with what you have. However, adding users to the home network and devices that can stream content over the Internet is another story. If the tested speed is less than 3Mbps down and 1Mbps up, you may not be able to enjoy the wealth of services available to higher speed customers, including streaming video services such as Netflix and VoIP phone service. While you may be able to have more than one thing going at a time, slower speeds are just not going to cut it with lots of things happening simultaneously. Just keeping things up-to-date, with the latest software updates, higher speed is becoming a must. Update downloads are routinely hundreds of megabytes, with some updates over 4 GB! It sure is nice to be able to pull down whatever it may be in less than 30 minutes instead of several hours, even overnight, for just one file. My advice to those interested in having the most fun and being the most productive is to get the highest affordable Internet speed. **Society is moving more toward online activity, so it is ever more important to have the capability for higher speed Internet.**

For now, let us assume such high speed Internet service is available *and* affordable.

Modern routers with both "n" wireless technology, and compatibility with the older "b" and "g," and four-gigabit Ethernet ports are available for as little as about $60 new. There are numerous brands available, and because of this, it is difficult to wade through the choices. If you want to read *reliable* reviews, ALWAYS check to see what **PC Magazine** has to say, online at **PCMag.com**, along with common sense and the advice from friends who have been there, done that. While the PCMag.com reviews may not be complete in that not all relevant brands and choices are always included, the source remains a good one to review. The caveat with regard to any review is this: in product roundups be sure brands in which you have interest are considered.

Wireless router brands to consider include **Linksys**, **D-Link**, **Belkin**, **Netgear**, **TrendNet**, **Buffalo** and **Apple**, in no particular order. For several years I have used **Linksys** and **D-Link**, as well as **Netgear**. Currently at work here at *Gadget Central* is a now discontinued **Linksys wrt320n**, occasionally still available online as a *factory refurbished model* – search for **refurbished wrt320n** and see what pops up. This router meets the criteria above and has performed admirably, with adequate wireless range to cover all of *Gadget Central*. In critical areas here, we are wired for Ethernet connectivity to take advantage of maximum speed while preserving wireless speed and availability at full throttle, as well. As with other electronic items, sometimes they just die. This is NOT uncommon with routers, either. During preparation of this chapter, my wrt320n gave up the ghost. Upload speed became unreliable. Through some trials, it was determined that it would be best to just get a new router.

I've just implemented a top-level dual band **Linksys E-4200 router (linksys.com), under $120 for a factory-refurbished model and about $160 for a new one.** Shop online, of course! This model is fully compatible with devices that are both 802.11b/g and 802.11n, on *separate channels*, to allegedly achieve maximum throughput (speed) on both wireless channels. "g" and "n" devices operate on different frequencies and with different specifications. By having this dual-channel capability not found on many other top brands that are easy to set up by the consumer, "b/g" and "n" devices do not have to share a channel, which can slow things down to the lowest common denominator. The disadvantage of using a dual channel setup occurs if you want to connect and share data across all devices. In this scheme, "n" devices cannot jump across to share info with "b/g" products on the other side. I probably will not use this capability because I want to share across and among lots of "g" and "n" devices. If you are seeing stars after reading the above, you may need to seek the advice of a geek, or search online for information and education.

The router is performing well, with a bit more wireless range than the old one and otherwise exemplary performance.

While this router was simple to set up, it is NOT the only one that is easy to set up and use. I've heard wonderful things about the little-known-to-consumers TrendNet routers, such as their **TEW-634GRU** just now found online for about $66. It has n-speed wireless, four gigabit Ethernet ports and, I am told, is easy to set up. Most users would not need a more expensive model. There is NO one brand or model that is best for every user, which is so very confusing.

I wish there was a way to help us all by recommending just one router make and model and being done with it. I regret to report that this is not the case. I do hope, however, that I have helped each reader to believe in the concept that the most expensive router, for example, is not necessarily the best.

If there is no economical way to bring Internet to your TV area via Ethernet, there are other solutions. **One I like and have installed with complete success is from** Netgear (netgear.com), **their** four-port 3DHD WIRELESS HOME THEATER NETWORKING KIT WNHDB3004. **Found online for** $200, it is much less than the cost in many instances of a hard-wired Ethernet connection. **The price may seem steep, but the** beauty of the product is that it is like a switch, plug and play and forget it. That's all I did and it just worked! Capable of sending and receiving at speeds *rated up to 300Mbps*, it installed seamlessly. In the home in which I did the installation, there is a Panasonic TV and a Panasonic Blu-ray player, plus a Dish Network satellite set-top box. The equipment had been hooked up many months ago without Internet connectivity. This method and using this or any similar product is not at the full speed claimed nor available if using a wired Ethernet connection directly from the gigabit Ethernet-equipped router, but it should suffice. For the record, wireless speed on devices such as this is rarely, if ever, as good as the stated theoretical maximums, and may not be even close. Regardless of measurements, *if it works, you're good to go*, but check it to see what the real numbers are just so you'll know. The easiest way is to Ethernet-connect a laptop to one of the ports on the end nearest the TV and other equipment and then run the test using speedtest.net. Be sure to temporarily disable wireless on the laptop for the testing. If it does not work, it might be that the distance is too great or that there is interference. In my test installation, the distance is line-of-sight about 20 feet from the router without obstructions between the two.

The Panasonic plasma TV for which I installed the Netgear device is equipped with Panasonic's Viera Cast, a series of apps built into Viera Cast-equipped Panasonic TVs. As soon as the cables were connected, I saw on the screen that a firmware upgrade was available for both the Blu-ray player and the TV! Both were completed in short order. The most amazing part of the installation was that Viera Cast *lit up* with numerous apps. My friend was astounded, to put it mildly. She liked her TV and Blu-ray player, both of which I had recommended months earlier. Now, she *LOVES* her TV and Blu-ray player.

Other similar plug-and-play solutions are available from other manufacturers, but you would be hard pressed to find any that are easier or higher performing for that price. Also check out similar products from Monster Cable (monstercable.com).

Powerline network adapters also exist with similar speed specifications and ease of installation, most with single Ethernet connectivity. In these installations, a switch such as the TrendNet above would be required to accommodate connection to additional pieces of nearby equipment. Some are supplied with a four-port switch as part of the kit. Makers of powerline network adapters include TrendNet, Netgear, ReadyNet, and ZyXel, among others. These network adapters send signal over *home wiring*.

This solution has several *"gotchas"* that need to be sorted out. If there is a split electrical panel (without a bridge from one to the other) at the home where a powerline network adapter is considered, these adapters *will not work*.

These adapters cannot be used when plugged into a surge protector strip, which is a must for all AV and other electronics installations, so they need to plug in directly to a wall outlet. Surge protectors strip out the signal.

They cannot be used in some instances in which there is too much line noise that will defeat the signal. Interference (line noise) from vacuum cleaners and hair dryers could negatively affect the performance of powerline network adapters. These products could also interfere with lighting systems that have a dimmer switch and short wave radios. And they cannot be used simultaneously with most if not all wireless telephone extenders that operate in similar fashion to these powerline devices, putting a phone jack where none exists.

With powerline Ethernet extenders, Ethernet at the router is plugged into a port on one of the two supplied extender modules, both of which plug into a wall outlet for power. The other module, at the TV area, uses another Ethernet cable plugged in, the other end of which plugs into a multiport switch and then on to the different pieces of equipment.

Had the wireless solution not worked in this test, I would have tried one of the powerline network adapters, most likely the ReadyNet E200K HD Network Adapter (Two Pack).

Surge protector strips are a must in any installation in which there is a computer, printer, TV, A/V equipment and other consumer electronics products. They are **NOT** alike, and some are no more than fancy extension cords. Reliable surge suppressor/protector products need to be able to stop a spike of electricity that can and would otherwise take out your equipment. Once a hit is detected, these devices instantly turn off, in a *nanosecond*, ideally even before the "hit" makes its way through to the connected equipment. This is very fast, indeed. Inside is a fuse-like mechanism that can be "blown" if hit with a significant jolt, to prevent plugged-in products from being effected. In such instances, the surge suppressor strip "sacrifices" itself so your equipment can survive and live on. Surges can come from the outside *and* from inside the home, from such things as refrigerators and washing machines, and more.

For more than 25 years, I have used and recommended Panamax (www.panamax.com) products for this purpose. I've had two hits that have sacrificed a strip, one of which also took out a connected product. It was a long time ago. A fax machine was toasted, but the Panamax warranty that comes with their devices kicked in and Panamax replaced both their device and the fax machine! This is the protection we all need. Look for devices that shut down and reset in the event of over-voltage (not just a surge) as well as prolonged *under* voltage, sometimes called a *brown out*. All my original Panamax strips except the two aforementioned devices are still in use and still protecting equipment throughout *Gadget Central.*

Another reliable and recommended brand is Tripp-Lite (tripplite.com). I also use and continue to recommend this fine, reliable brand of not only surge suppressing products, but also battery back-up products, a subject for another chapter! Stick with these brands and you cannot go wrong. Find the right product online for your audio/video installation, as well as for laptop computer chargers, computers and related equipment.

ALL flat panel TV screens need to be cleaned carefully. Use ONLY a clean and freshly washed microfiber cloth, preferably a large one. They are inexpensive and you might already have and use one or more. **NEVER use paper towels.** **NEVER** use fabric softener in washing or machine drying a microfiber cloth.

DO NOT EVER use a chemical glass cleaner or any other cleaning chemicals, such as Windex, on your TV or other display surfaces such as a computer monitor. DO USE plain water, preferably filtered and even better, distilled. Never put water *directly* on the screen or anywhere else on a TV or on **any** electronic equipment.

Use a fine spray-capable sprayer bottle. If you have one not previously used for something that cannot be thoroughly cleaned out, get a new one. Be sure to spray some plain water through a used spray nozzle with its clean bottle, and then fill the bottle with whatever water you have chosen, and make several sprays into the sink to get it primed and using the "new" water. New bottles are inexpensive and readily available from many stores.

I use and recommend **Tip N Spray** bottles, uniquely designed to allow delivery of nearly every drop of the bottle's contents. No more "dry" firing sprays. Check them out at **tipnspray.com** to see their "Aha!" video demo and for retail availability.

FIRST, though, get and use a can of "air," such as **Dust-Off** (**dust-off.com**) to blow the dust from around the TV, perhaps along the top and in cracks and vents on the side and/or rear of the set, as well as along the bottom of the screen. **Read and follow directions carefully, please.** These sprays can be dangerous if proper use and care guidelines are not followed.

Then, use the freshly washed microfiber cloth, lightly misted with water sprayed directly onto that cloth to make it only *slightly* damp, and only on the surface area to be used to clean the screen. Now, wipe the screen, gently, please. That's it. Water is FREE. We like FREE! Any fingerprints or other *schmutz* should come right off. If additional cleaning capability is needed, use a few drops of rubbing alcohol on the same lightly damp cloth and gently clean the stubborn spot(s). Not much sticks to the TV's glass, as you will notice over time. THEN, *after* cleaning the screen, wipe the other area as needed. I also occasionally use a previously unused-for-anything-but-dusting horsehair paintbrush to get into slots and other areas. Always use the microfiber cloth on the screen first, before other areas on the TV (or computer display). You see my logic here? When wiping other areas after the screen, don't worry if you had put a few drops of rubbing alcohol onto the microfiber cloth as it quickly dissipates. Discourage body parts or anything else from touching the screen. A clean screen is a distinct pleasure to watch.

CHAPTER 2

RADIO

REDISCOVER RADIO!
CONVENTIONAL AND INTERNET RADIOS

I am an inveterate radio listener. I listen in the car, at home in various rooms, at bedside and in the shower, in my office and even while traveling on the road far from home, in *and* out of a car.

More than just music, I also like to listen to news and talk stations. Another favorite pastime is listening to old time radio (OTR), with shows from radio's golden age starting in the 1930s, all the way through to the mid-1970s when most radio dramas and comedies faded from the airwaves.

I am amazed when I hear today's radio hosts and radio ads reference the target audience said to be listening *in the car*. It is as if stations assume this to be the case. Have they forgotten that traditional radios operate just about everywhere? Shouldn't they promote listening, oh, I don't know, at home? In the office? ANYWHERE and EVERYWHERE? That's the beauty of radio – it's everywhere!

Baby boomers such as I grew up with radio *and* TV, and we were accustomed to listening at home more than in cars. In those years, commute times were so much shorter.

Everyone is encouraged to rediscover radio! Radio is still out there, and in many ways it is better than ever. *One particular advantage is that it requires only our ears (and part of our minds) to participate.* **All over-the-air programming is FREE**, as is much of the content streaming over the Internet. Contrast this with TV watching that requires sitting, watching *and* listening. Cruising the Internet is generally more visual- than audio-based. Listening to radio allows us to move about freely, using vision and part of our brain for other things. Try watching TV with your eyes closed! But, radio listening, on the other hand, is a pleasure with one's eyes closed.

Radio is also THE thing one would want to have on hand and available in the event of an emergency. In the absence of TV, which might not be available during a power outage, battery-operated radios work perfectly and provide the vital info link to news of the event and other crucial information.

Today's availability of radios is quite good. I particularly like those with outstanding battery life, to about 250 hours, high sensitivity to bring in weaker and distant stations, and with excellent sound, especially when it comes to optimization for the spoken word as well as very good music reproduction.

My favorites and those I recommend include standouts from the **C. Crane Company**, with all the info at **ccradio.com**. Once there, click on the link to **Radios**.

First up, the **CCRadio-EP**, $70, a fine and basic model with simple operation. Old-school reliable dial tuning, AM, FM, with treble and bass controls, switchable lighted dial and with proprietary AM tuning technology providing the best sensitivity of any portable radio regardless of price. It provides at least 250 hours of operation from four "D" batteries, and it comes with an AC adapter.

Next, **CCRadio-2**, $160, is also portable, powered by four "D" batteries and comes with an AC adapter. This model features precise all digital tuning with five memory presets per band and an excellent lighted display with large numerals and text, and a sleep timer from 15 – 120 minutes. This powerhouse receives AM/FM/Weather band/2-Meter VHF (or HAM) band. I know that AM and FM are obvious in what they do.

The Weather band encourages listeners to be informed with regionally localized reports from the National Weather Service. The 2-Meter band is the set of specific frequencies that are in use during national, regional or local emergencies, with information provided by volunteers with a love of amateur radio. This well-established and long-standing radio capability was notably put to good use in the US during the Hurricane Katrina disaster along the Gulf Coast. Dedicated HAM radio operators were among the first to report on what was happening locally and provided valuable information to citizens and emergency responders alike. With this capability inside the CCRadio-2, owners are equipped and prepared to receive info and to listen in on conversations locally, up to about 100 miles away, and sometimes beyond. (More information about the 2-meter HAM band is at **bit.ly/rqc0sN** – link shortened for your convenience.) CCRadio-2 is an outstanding performer in all regards, and also has super tuning capability as well as excellent sound, especially optimized for the human voice. It is a perfect bedside "clock radio."

The radio I take along on all my travels is the $50 **CC SWPocket**. Hand-sized and battery or AC operated (adapter included), this model digitally tunes in AM/FM/Short Wave, plus it has a clock, alarm and sleep timer. Operates up to 70 hours on a pair of AA batteries! The best performing travel radio, even among those costing up to $200.

In MY shower, and capable anywhere wet or dry, is the **Sangean H201 AM/FM**, about $60 at **amazon.com. This is a radio created from a concept by and with features requested by** *The One and Only Mr. Gadget*®**!** I asked for it and Sangean (**www.sangean.com/first/first.asp**) built it. Enter the URL above and then click **Product**, then **Special Application Radio**. It is without peer as the best at what it does, if I do say so myself. This radio delivers great sound, even when wet, and provides about 250 hours of operation from a pair of "D" alkaline batteries, five presets for each band, with timer AND auto off from 10 to 60 minutes. There is a side-aimed LED flashlight ideal for power outages and other uses, and with an attention-getting HELP alarm sound, too. The H201 radio comes with a wall mount plate and built-in rotating handle hanger. I use dots of silicone sealer to mount the plate on the tile wall in my shower. The supplied adhesive strips don't work well. Use this radio wherever there is the need for a weatherproof (not waterproof) radio – at home, near the pool, on a boat and at the beach, though I would not recommend letting it live in a sandy environment.

Ideal for the kitchen area and elsewhere is another Sangean radio created from a request by and with features asked for by yours truly. The AM/FM **K-200** features a vertical, space-saving design, with top-mounted waterproof membrane buttons for power, station presets, an egg timer, voice memos, and more. It includes a down-facing speaker, with night-light, clock, two alarms, digital display and more. The K-200 has very good sound and long battery life from four "C" batteries or it can be operated using the included AC adapter. It is available in pink or white and best priced from **amazon.com** for about $67 - $76, depending upon color.

By the way, I have received not one penny from Sangean for my ideas and suggestions! I am told the H201 remains among their best sellers.

While these are great radios, as standard radios go, **I am also a BIG fan of Internet radios**. These are devices that *look* like standard radios but they are really computers, connecting to home or office Internet networks, wired using Ethernet or over Wi-Fi. Ideally for home or office use, these "radios" *receive* radio stations through their Internet streams. Most broadcast stations also make their programming available online, which means Internet radios can receive and stream the stations, both music and talk. Wherever in the world may be your hometown or favorite place, chances are good that radio stations from there can be received by an Internet radio. Be *here* and listen as if you are *there*, wherever *there* may be. That's over 18,000 stations! Many of these devices can do much more.

For anyone with broadband Internet connectivity and poor standard radio reception, Internet radio is a welcomed and perfect solution! But, there is so much more to Internet radio than just interference-free reception of conventional radio stations.

In addition to so many interesting music stations, I like to hear news from the world news service of the BBC, local news in the UK and from other English-speaking countries, including Canada. Are you a native of a non-English speaking country? Listen to radio stations from virtually any country on your own Internet radio!

I get all my local Los Angeles stations and listen to local stations from many other cities I've visited in my travels.

As I indicated above, I love to listen to **old time radio**. There were hundreds of shows, maybe thousands. Many of the best and most notable are still available over the Internet, so they are available on Internet radios, with many available at no cost, just as are all the local radio stations. Some programming is available to subscribers at services such as Live365 (**live365.com**), for a fee as low as $5.95 per month with an annual plan that provides unlimited access to all stations and not just old time radio. Many old time radio stations are available from live365.com at no charge, too. As well, there are numerous music channels to satisfy nearly any interest.

Additionally, there are many other free and paid subscription services providing every kind of music imaginable with compatibility through many of the Internet radios.

The ONLY Internet radios I recommend and use are also the easiest to use and set up and with the greatest capabilities. **Those from Logitech are clearly the best.** If you look up the subject online – **Internet radios** – you will find so many brands and choices that you will be amazed, dazed and confused. Take my advice and stick with my simple suggestions. Visit **logitech.com** and click to see **Audio** and then **WiFi Music Players**. Shop online.

Quickly stated, the others not recommended are often more expensive, more difficult to use and lack useful capabilities of the recommended brands and models.

Recommended Logitech models include **Squeezebox Radio**, about $165, and **Squeezebox Touch**, about $236, both from **amazon.com**. Squeezebox Touch does not have built in speakers. I connect this one next to my desktop computer, sending sound to a pair of excellent Bose computer speakers that feature dual inputs. You might consider this model to live with your existing A/V receiver, connected through it to the receiver's speakers. Use the excellent supplied remote control to access programming. Get best prices shopping online.

Look up these models to see capabilities much greater and easier to access than all others. **Squeezebox Radio** has all the typical clock radio features, and with features unheard of in a conventional radio – *multiple* individual alarms with wake-to capabilities including radio stations and myriad built-in natural sounds and sound effects, Facebook and news feeds on the sharp, color screen WITH background sound from Internet stations, to name just a few interesting features. Oh, and display brightness is widely adjustable in both ON and OFF modes. Seriously, **DO look them up at the Logitech Website to see all the features and almost unimaginable capabilities**.

Logitech Internet radios are set up **online** by the user through the **mysqueezebox.com** online portal, with YOUR chosen apps and settings, tailored to YOUR interests. Each radio has its own unique and individually addressable Internet number, called a MAC address. I set up my preferred stations under *Favorites* in the order of my choosing, and regardless of their source, along with all the other applications in which I have interest, all online. The info is automatically, instantly mimicked on each addressed plugged-in Squeezebox device.

A basic, simpler Internet radio, without the extensive and robust feature set of the Logitech products, is the highly recommended **CC WiFi Internet Radio** from my friends at the **C. Crane Company – ccradio.com**. This $140 powerhouse tunes in Internet audio streams of radio stations across the world as well as user accounts of **Pandora, Aupeo, Live365** and **MP3Tunes** (look them all up online). Other features including sleep timer and alarm. Its provided remote control accesses all the features on the pleasant, backlit digital display and its small footprint allows the **CC Wi-Fi Internet Radio** to fit almost anywhere.

Unlike the portable radios recommended above that all accept batteries, and can also be powered by plugging into the wall, Internet radios are not designed for operation on battery power. They need too much *juice*, more than batteries can supply for many hours of operation. You will see, however, that there is an accessory battery pack available for the Squeezebox Radio, but it is not designed for long use away from power.

If you become a serious Internet radio listener and wish for service to be available during a power outage, at least for a few hours, be sure your Internet equipment is plugged into a battery backup unit, as detailed in the chapter on **Home Phone Service**. Battery backup is also required for an Internet radio to be online in a power outage. Remember, this "radio" is really a little computer, and needs power and Internet connectivity to function, whether using Wi-Fi or Ethernet connection to the router. **Think about this if the Internet radio will be relied upon as a wake-up alarm** if the area in which you live is prone to power interruptions. If the power is out and the radio is not supplied with AC power, the alarm cannot function. My battery backup at bedside and the one powering my Internet-related equipment assure that, barring a major catastrophe, my bedside Internet radio will operate as normal, and any alarms that are set will also continue to function.

For those who like using a bedside radio of any sort and who also have a sleep partner, I recommend, use, and endorse (with NO compensation), the $20 **SoftSpeaker Pillow Speaker** from **C. Crane Company** (link shortened for your convenience - **bit.ly/t15RMx**). This little speaker plugs into any radio's earphone jack. I use it so I can listen at night while *Mrs. Gadget* may be reading or sleeping, or so I do not disturb her enjoyment of a different station on her radio. It's just about perfect, and better than anything else on the market. My only item on the wish list for this product would be for better jacketing on the wire from speaker to the jack on the opposite end. Over time, the little bit of heat generated by the speaker doing is thing takes a toll on the end of the wire at the speaker. It becomes less and less pliable and stiffens. I've not had one break as yet, but I wish there was a better quality, always-pliable cover over the wires.

The folks at the C. Crane Company have designed-in smart ideas, one of which I found from only one other radio maker. When an earphone, headphones or pillow speaker is plugged into their CCRadio models, and if an alarm is set, their intelligent radio design dictates that the alarm sound is sent through the radios' *internal speaker* and *not to whatever is plugged in at the earphone jack*. How smart is that! Think about it. Users would not want alarms to be piped through headphones or a pillow speaker, as this would not awaken the user. Check any of your current radios, and in particular your bedside radio, to determine how your radio responds. This smart design is also present on Logitech's recommended **Squeezebox Radio**, mentioned above. The **SoftSpeaker Pillow Speaker** from C. Crane is the ideal companion for these radios! The cord is long enough for easy reach from the radio to a nearby pillow. Depending upon the radio I am using, with some having stronger output to the earphone jack than others, I prefer placing the pillow speaker under my pillow. Some users prefer placing it on top of the pillow with an ear resting directly on the speaker's soft outer cover.

Search *Internet radio* on **MrGadget.com** for reviews.

Another worthwhile and FREE SERVICE that is ideal for Internet radio users is **DAR.fm** (dar.fm). DAR stands for Digital Audio Recorder. I use their FREE service to record favorite radio shows so I can listen to them at a more convenient time. For example, when traveling from my home in the west to the east coast, the three-hour time difference means I need to get to sleep at a time that is three hours earlier on the west coast. Instead of missing a nighttime show, I set my free DAR.fm account to record what I would otherwise miss. Then, I can listen at my convenience using a FREE **MP3Tunes** account (**mp3tunes.com**) on a computer, Internet radio recommended above or through the FREE mp3tunes app on an iPhone, iPod touch or iPad, plus Android phones and **Roku** boxes (**roku.com**). And I can also use MP3Tunes to upload and stream my own music to any of these devices. How about that!

If you own a smartphone or tablet powered by Android, Windows or the Apple's iOS, there are downloadable apps allowing users to tap into these providers and stations! Search under *Internet Radio* to find them at your device's apps store. I like the FREE TuneInRadio (tunein.com) and IHeartRadio (iheartradio.com) apps for mobile devices. I also use and recommend the $4.99 OoTunes app (**ootunes.com**) for iPhone and iPad. These apps are power and bandwidth hungry because content is streamed, so they are best accessed with external power supplied and using Wi-Fi instead of data from the phone's monthly data plan, unless unlimited. Computers also easily access Internet radio content. When traveling, I listen to home and other stations of interest, as if at home, on a laptop, smartphone or tablet. I feel connected to home while traveling as I listen to these same favorite local, distant and special interest Internet-only stations on my devices.

CHAPTER 3

COMPUTERS

LAPTOP OR DESKTOP, WINDOWS OR MAC, OR MAYBE A TABLET?

Once upon a time, desktop computers had the clear edge in price and performance.

What do most of us do with our computers? Email, Facebook, Web surf, watch funny (or disturbing) YouTube videos that are forwarded to and from everyone we know, perhaps manage family photos and even a music collection. Maybe we do a bit of word processing, communicate using audio and video with far-flung family and friends using Skype, which necessitates the use of a webcam. Some keep track of investments and handle buying and selling on eBay, as well as other online shopping and buying. Some users also play simple games.

Still others may use their computers to edit home videos and post them to YouTube or elsewhere. Photo enthusiasts frequently use computing power to do wonderful and creative things with digital photos. These both require more than base-level computing power.

Today, anything short of the serious gamer would do perfectly well with a laptop computer. Laptops offer the ultimate in portability and at surprisingly modest cost as compared with offerings from just a few years ago. But, **it's**

the portability of a laptop that is so freeing and pleasant, as well as just plain useful. Connecting to the Internet is just as easy with a laptop as with a desktop computer. **I can think of no typical user today that would be best served with a desktop computer instead of a laptop.**

Whether using a desktop or laptop, Windows PC or Mac, I recommend that everyone use a router, a device that connects the computer to the Internet, either through a separate modem or through a device that combines the modem and router. Routers work in concert with all wired forms of Internet access, such as DSL and cable systems, as well as with fiber optic systems including Verizon FiOS. The router provides what is called a hardware firewall, a barrier between the bad guys out there and your computer. No, this is not to stop Windows-based malware, but to prevent unauthorized access into your computers or onto your network.

Many Internet service providers offer a combination device, a modem and router in one that will be just fine for most users. HOWEVER (you just knew I'd have one of these), this is not always the case. It all is dependent upon the needs of the user.

If the user is not so *into* downloading, uploading, transferring data from one device to others, does not share the Internet with others in the home, does not stream video or does not experience slowdowns the way things are now, go with what the provider offers.

Most cable companies charge a monthly *rental* fee for a modem. Buy your own and save! The break-even point will vary, obviously. Let's say your provider charges *only* $3 per month for the modem. That's $36 per year and after only five years, your investment in that rental modem is a whopping $180, and the charges add up after that, too. If your cable company charges more, you're really on the hook! I visited a friend recently and asked to examine her cable bill. I saw a monthly modem rental fee of $7 PLUS a monthly wiring insurance fee of $4. Outrageous!

You see, I remembered that nearly 10 years ago I purchased a modem for her and installed it, returning the cable company's modem to stop that monthly charge! And she had not authorized the wiring insurance either. Fortunately, we learned that the modem rental had "accidentally" been applied for just under two years. Our initial call to the company yielded an offer of only a courtesy refund for three months on the modem fee. I think NOT! We asked for a supervisor who had no reason to not agree that the charge was not justified and provided a complete refund for all modem charges, and cancelled the monthly wiring insurance fee. Unless the residence is prone to rodents eating in-wall wires, the wiring should remain intact for more years than most are in a residence. The insurance ONLY applies to inside wiring. Any cable company wiring beyond the residence walls is the responsibility of the Internet provider and not the customer. This is the standard everywhere

Buy your own cable modem and don't forget to return your rental. Look online for the cable modem you now have, provided by the cable company. You know that one will work! OR ask your provider if the modem required is DOCSIS 2 or DOCSIS 3 compliant. These are industry standards. Expect online prices to run about $50 for a DOCSIS 2 modem and still under $100 for a fancy DOCSIS 3 modem. Get the type used by your provider, regardless of brand. Once you have modem in hand, you will need to provide the individually addressable MAC number from the new modem to your provider so it can be entered into the system, enabling THAT modem for YOUR account. They rarely fail, so it's a good bet you will save money every month from then on.

If you have Internet service at home or in a small office, you NEED a router, too. I'll explain shortly.

Most phone companies do not charge a monthly fee for modem or router. The idea is to not incur such a fee for this equipment.

If you're hankering for more, feeling the need for speed when moving quantities of data around within your network, or experience other than stellar performance, a new, better router than the one from your Internet service provider as well as faster Internet service should be on your wish list! It is the Internet provider that determines overall speed, but speedy Internet and a good-performing router allows users to realize the faster data transfers not only online but within the home network and between devices.

Oh, and if you have a smart phone and use it at home, you have to be nuts to be without a wireless router. Why, oh why would anyone NOT use FREE Wi-Fi through a home router rather than use precious, high-priced data services of a mobile plan? I have an old friend who recently told me that she burned through her new smartphone's data plan allotment in one week while at home and was surprised at the added cost to her bill. She had NO router! That was a costly omission. Now, she knows how to have her smartphone fun AND not pay an arm and a leg to use it from home.

Back to the original recommendation that everyone use a router. If without one, get one, simple as that. Even a simple router without wireless capability, connected to a single stationary desktop computer, is mandatory for safety's sake. The router performs as a powerful closed door, so anyone trying to get into your computer from the outside will be stopped at the router. YOUR computer becomes invisible to outsiders in this way. A router will NOT stop malicious software, spyware or viruses from entering your connected computer, because you "invite them in" through email or by visiting certain Websites.

Wireless **routers** are inexpensive, as little as about $50 online for a basic wireless "n" model with four ports of 10/100 (speed) Ethernet, so there is no reason to get a wired-only model. Wireless routers also allow multiple *authorized* (by you) users to share a single Internet connection. In this way, mom and dad, all the kids, and even visiting grandkids plus authorized visiting friends can share that one Internet connection wirelessly, as well as connect through up to four wired ports on the router. Recommended wireless router brands include **Linksys**, **D-Link**, **Netgear**, **Belkin**, **TrendNet**, **Buffalo** and **Apple**, among many others. I am using a fancy, excellent-performing **Linksys E-4200**, though others, such as a **TrendNet TEW-634GRU** at about $66 should do a fine job for most readers, with easy setup.

What about screen size? Laptops with up to 15-inch screens are easily portable. Those with larger screens are hardly, *easily* portable as desktop computer replacements. A laptop computer's screen may provide enough real estate to satisfy your need, but if you need more, the addition of an external screen is inexpensive.

Desktop-size monitors, the kind that are of sufficient quality for most of us, are quite inexpensive, too. Most users would want, say, a 24-inch monitor for a better, bigger view than what can be found on a laptop computer. This is the "sweet spot" in size and price today. For as little as $125 and up, perfectly acceptable monitors in this size are available. Higher cost, better quality monitors are targeting users who routinely work on graphics or have need for precision when manipulating photos. Few who are reading this would benefit from more than what is recommended here.

Some users prefer a 30-inch monitor, and even those are only about $300-$400 for good, though not professional quality. Shop online for the best deals and read reviews to find a consensus on quality within the brand considered. Features to consider include highest contrast ratio and lowest refresh rate, also called response time. I don't have a strong feeling about best brands of basic monitors, but traditionally, ViewSonic, HP, Dell, Asus, and Samsung, among others, are the leading names. I use, enjoy and recommend a high resolution Dell UltraSharp 24-inch model (the one I use), but I am eyeing a 27-inch for upgrade. I'd buy another Dell monitor without hesitation.

If using an external monitor, laptop users may also benefit from a connected full-size keyboard and external mouse. This is because the laptop is usually placed in front of the monitor, between it and the user. In order to take full advantage of the connected external monitor's real estate, the laptop must be in a position so as not to block the view of that larger screen. So, either the laptop needs to be nearly closed or off to the side. In either case, access to the laptop's keyboard is thwarted or, at the very least, not in an ideal position when used with an external monitor.

External devices for keyboard entry and mousing become necessary. Look to Logitech for an external keyboard as well as one of their excellent wireless mice with extended battery life, at least one of which is rated at 36 months. Mac users are advised to look at the Logitech Wireless Solar Keyboard K750 or choose the excellent $49 Apple Keyboard with Numeric Keypad. A wireless keyboard is not needed here, in my view, and only adds to maintenance requirements. Another useful Mac accessory is Apple's $69 Magic Trackpad, duplicating ALL the gestures capabilities of your MacBook Pro! This is available only as a wireless accessory.

A final keyboard advisory for Mac users – Get a keyboard cover! I recommend against having food or any spillable liquids within the dangerous vicinity of any computer. Mac users have an easier time finding keyboard covers because of the uniformity of products. There are but a few Mac models for which to design keyboard covers. The ONLY Mac keyboard cover I use and recommend, based upon extensive testing that began long ago is a **KB Covers** model of your choosing from **KBCovers.com**. Protecting against dirt and dust, as well as other airborne matter flying about, they are worth the cost just because they keep all that junk from contaminating the guts of the keyboard. Inside any keyboard can become really disgusting without one of these! Find many specialty and "plain" covers to your liking, but get a KB Cover for your Mac AND for other Macs for which you may have responsibility, such as student Macs. Priced at $25 for the non-specialty models, they fit and feel best, they protect best, and they carry a lifetime warranty. I've trusted them for several years. Unfortunately, there is no recommendation available for the myriad Windows computer keyboards.

When in need of a **Webcam,** use the one built-in on the laptop instead of having to buy one for an external monitor. If Webcam use is going to be frequent and if the laptop is going to be used with an external monitor most of the time, do get and use an external Webcam for its added convenience. If there is a need for a separate Webcam, **Logitech** is my choice above all others. The company offers an outstanding $100 (list price) model **C910,** ideal for Windows and Mac computers. This is the one I use. See it on the Logitech Website (**logitech.com**), and shop online for best price, usually about $60. Less expensive Webcams are available, down to about $20, but as with most products, you get what you pay for.

How does one choose a laptop computer? First, think about how it will be used. With prices as they are, I am not recommending a *netbook,* the little and inexpensive models that have 10- to 12-inch screens. Cost is about $200 - $300. As a primary computer, netbooks lack horsepower, do not have a built-in optical drive for DVDs and CDs, and the keyboard is usually smaller than full-size, for starters. Most are flimsily made, too. If budget is a primary concern, then I suppose I can bend my recommendation to include these; however I'd rather not, in favor of greater satisfaction over at least three years use with a better computer. If all you want or need is a basic laptop, get one with a 13- or 14-inch screen. In these sizes, it is also easy to find models with as-good-as-desktop-computer performance, and even better than some desktop computers!

Now, the time has come for the discussion of **Windows versus Mac.** This topic brings out the fangs in rabid fans with opposing viewpoints.

The war is over! Windows 7 on a PC and Mac OS X on Apple computers will both get the job done. Yes, Macs cost more, from a little to a bunch, than *most* Windows laptops and desktops, even similarly configured. Get what you like. If the computer is just an appliance, a means to an end, then Windows will satisfy.

That's great advice for YOU, but not for me.

I believe the world of personal computer users would be best served if the overwhelming majority used Macs. Yes, the difference between the Mac OS and Windows is not as great as it once was. A Windows computer usually starts life doing fine, but in about one year, sometimes more, sometimes less, *something* often goes awry, either with the Windows operating system, with an application or due to a virus or other malware.

I know that Mac users are usually happy campers for four, five, six, even seven years after purchase. Though it is possible for there to be a virus or other malware that can affect Macs, this just has not yet happened (knock wood or do some other superstition-based acknowledgment of your choice). Still, I recommend every Mac user install a Mac antivirus program, including Intego **Virus Barrier X6 (intego.com)** or **ESET Cybersecurity for Mac (eset.com)**. I'd rather pay for it and NEVER need it than not have it and face disaster if it should come to pass that a Mac virus is in the wild! (NOTE: I also recommend Intego's **Internet Security Barrier for Mac**, with its additional automated Mac back-up utility, covered in the chapter on **Computer Backup**.) A FREE Mac antivirus program is also available called **Sophos (sophos.com**, then **Products**, then **Free Tools).**

The only downside to a Mac, if there is one, is that Mac laptops cost more than many Windows laptops, but there may not be such great disparity with laptops similar in performance and quality. Even a basic, $1,199 13-inch MacBook Pro is going to perform better than many Windows laptops. Performance is relative, too. It is not just how fast, but how reliable, how easy it is to use and the all-important fun factor. Macs also come with everything needed to get started, including photo, music, and video applications, mail, calendar/appointments, and more, all of which are nicely integrated with the operating system and play nicely with everything else on the Mac. Instead of tinkering to get things to work at all or to work correctly, or working out a problem, Mac users traditionally enjoy the experience for many years without fuss.

I've not included the $999 MacBook Air here, as it is not a good primary computer *for most users*. There is limited internal storage space due to there being a lower capacity than traditional rotating media in its solid-state drive and there is no built-in optical drive. However, an external optical drive is available for $79. If this model meets your needs, it would be a fine choice.

As this is a favorite topic of mine, I have been through countless occasions in which others take my advice and start with or switch to a Mac. It is a fairly universal reaction by Mac switchers to exclaim that they should have done it sooner. **Never have I experienced a disappointed person who has regrets after switching to a Mac.**

I neither work for nor have any arrangement with Apple (or other companies, for that matter). As with everything else you will read, my opinions are based on my own experience.

When it comes to price, *the cost of the goods*, would you rather spend more up front in exchange for likely more than twice the time of satisfactory service? Sure, any computer, Windows or Mac, can experience a hard drive failure. And any computer can experience other maladies. However, it is my belief and experience that Macs do better for longer and that Mac owners are more satisfied, longer. Learn more at **apple.com/mac**.

Many detractors are anti-Apple, with its closed system and vice-like grip of control on hardware and software. However, there are numerous advantages to such a monopoly. From hardware to software, everything just *works*, and the simplicity of the Macintosh operating system, the Mac OS, is the glue that holds everything together. For those wanting Microsoft Office, there is a Mac version that is fully compatible with files sent to or received from Windows-based computers.

If Windows is still your choice, there are almost too many computers from which to choose. This is part of the problem, in my view. To a large extent they have become commoditized. The general rule is that the cheaper the Windows laptop, the less likely it is to last as long as may be desired, and to last without showing its age. I've had Windows PC laptops for testing purposes that develop shiny spots where my fingers go most often in as little as three months. The chassis and outer plastic easily shows signs of use on such models. On some I've had that are comparably priced to Macs, operating anomalies have developed in less than one year. Sure, I *might* be able to remedy the issue, but I don't want to have to bother with these things. I just want to compute and enjoy life when not having to be at the computer. Time is money!

It cannot be argued that the majority of personal computers in consumers' hands are running Microsoft Windows software. This does not mean Windows is best! It does mean that Microsoft and partners have been more successful in marketing efforts. In sales numbers, the last time I checked, the 13.3-inch MacBook Pro model was the second best selling laptop in 2011. Apple is a strong seller, gaining in market share worldwide!

What is recommended for Windows laptop shoppers? Be sure to choose a model featuring the newest Intel architecture. This means one with the second-generation Intel dual-core i5 or quad core i7 processors. The sweet spot is the i5 processor. I do not recommend models with a lesser-performing i3 processor as this chip will not provide a satisfactory computing experience over the long term. The best way to assure getting this dramatic performance advance from Intel is to buy the newest computer model that meets your needs. Current Windows models and Macs use these processors.

The *other* processor brand in Windows computers is from AMD. In the past, AMD and Intel jockeyed back and forth as the preferred processor. Today, however, Intel is arguably at the top of its game and offers the best performance and reliability for your computing dollar. If you differ and find an AMD-powered Windows PC you like, forget my advice and do as *you* wish. I won't come after you.

Which Windows PC models or computer makers are recommended? You should find what you want (in this order) from **Toshiba, Asus, Dell, Samsung, HP, Compaq, Sony, Lenovo, Gateway** and **Acer.** Shop online through the makers' Websites to find the model of choice. You may also wish to visit a retail location. I must warn that this is confusing, much more so than shopping for a Mac with just one brand and only a few configurations.

Some Windows PC retailers have models made only for them. Also, Windows models change with great frequency, much more regularly than do Macs. Windows computers are very close in configuration among models in a "family." "Deals" that may be found often omit features and upgraded hardware, such as a better graphics card, from certain models in order to keep a special, lower price OR there may be model deals that do not allow customization in order to keep a low price. Expect to pay from about $700 to $1500, depending on configuration for a Windows laptop that will prove reliable and perform admirably, at least to start. Many of the best prices are found at amazon.com, of course, but also check other online prices and prices at club stores.

For those still on the fence, **did you know that ALL current Apple computers can also, simultaneously if desired, run both the Mac OS *and* Windows operating systems?** It's what I do. My Mac is also a speedy Windows computer, running Windows 7. It is as modern and up-to-date as any native Windows PC I could want. I had to supply my own Windows 7 software and the few applications I wanted that are Windows-only, and, of course, antivirus software, but when I am running Microsoft Windows on my Mac, it is every bit a "Windows PC" as any true, out-of-the-box, Windows computer around. I have the best of both worlds in ONE computer with less cost than buying two computers. You can, too! Details are at all Apple retail stores. Some software is Windows only, which is why some of you may need this duality. If you don't *need* Windows, there is no reason other than "want" to run Windows.

Macs can be started up in their full Windows dressing, as a Windows-only computer. They can also use software that allows both operating systems to be running simultaneously, such as **Parallels** (**parallels.com**). Now, even as I am writing this volume, I have my Mac OS running AND, at the same time, Windows 7 through the excellent Parallels version 7 software. Doing so is seamless and also fun. The pleasant "eye candy" of windows moving about, dropping out of sight, sliding in and out of view is also an attraction, but this activity is also part of how Parallels integrates into the full Mac experience. **If you want or need Windows on your Mac, Parallels is for you!** Pertinent information about using Windows on your Mac from Apple's point of view is at **apple.com/findouthow/mac/#windowsmac**.

Finally, some readers may wonder whether a tablet computer would be a good choice *instead* of a laptop. If you do not need much onboard storage for music, photos or videos, are usually within reach of reliable Wi-Fi and can master the touch screen for all data entry, a tablet may be a good choice. Just remember that there is no optical drive for tablets, so all software or resident music must be Wi-Fi downloaded OR transferred wirelessly from someone else's connected computer. There may also be docks into which your tablet of choice can connect so as to allow a more traditional kind of keyboard and mouse to run as you use your tablet. At this writing, there are such devices for iPad. Search for the same for the tablet other than iPad you may be considering.

Tablets are fun to use and can be useful tools, as well. With prices starting at about $200 for full-function, Internet-capable tablets with color screens, they are an attractive alternative *for the right user, but not for everyone.* Tablet portability and functional capabilities as e-readers make the attraction even more worthy of investigation, so let's have a quick look!

Amazon's $199 Kindle Fire was a 2011 holiday season hit. Though with *only* a seven-inch color touchscreen and lacking cameras, it was the most talked about alternative to iPad in the realm of eReaders capable of Internet access via Wi-Fi. With memory limited to a maximum of 8GB, it pales in comparison to iPad's maximum 64GB. Ah, but the price contributed to its holiday success. Amazon Fire runs a version of the Android OS with Amazon's own custom browser called *Silk*. Ah, but the $199 price as compared with iPad 2's entry price of $499. Amazon indicated their users will store content at no cost on Amazon's cloud. Check it out at **amazon.com/kindlefire.**

Video looks better on Nook than on Kindle Fire. It's so nice to have choices.

If a dedicated e-reader is your forte and not an iPad or other tablet-type computer, also check out the **Kobo eReaders (kobo.com/ereaders)**, including THEIR newest, Kobo Vox, a $200 choice with a huge online bookstore and Android powered eReader.

Other choices abound, so it is worthwhile to do one's homework. More recent additions to the tablet computer space include speedier performance to rival iPad, but none yet rival overall ease-of-use, which is to be expected. Prices for most of these more worthy competitors are also closer to that of iPad's $499 entry price.

I can suggest no functional or other downside to Apple's industry-leading **iPad (apple.com/ipad)**. Only its expense that is more than many others is a factor. As with other Apple products, they just work, and iPad enjoys exemplary battery life. Cost aside, it is hard to find fault with iPad as the tablet most consumers would love to own.

The tablet computer field is huge and growing! Windows 7-based tablets are not strong players in this arena, with shorter battery life and no real advantage over others in any capability. 2012 may be a different story as Microsoft readies rumored Windows 8-based tablets for introduction. It will be an uphill battle.

Note that screen size is an important factor to many users. If primarily as an e-reader, the less expensive and smaller screen sizes of Kindle, Nook and Kobo devices, and competitive products can be just right. If e-reading takes a back seat to other interests, the more expensive, larger screen models will be more desirable, but the **lower price is an attractive argument that may overcome objections relating to screen size**! iPad makes an excellent e-reader, too, with Kindle, Nook, Kobo and other e-book compatibility.

CHAPTER 4

PRINTERS

STOP SPENDING A FORTUNE ON PRINTING & PRINTERS

I hear the same story with regularity – "Costs are SO high to use my printer that I barely use it!" "Even printing with just black, it *still* costs a fortune!" It hurts *me* to read and hear such sad tales, but I think all or at least most of it is entirely unnecessary.

The stories presuppose that the users have inkjet printers. I have to ask why. I have nothing against (color) inkjet printers, mind you. If that is your printer of choice, then so be it, but . .
.

Did you know there are printing solutions costing *less than one cent per page*? Ah, now do I have your attention?

It is admitted that most printed pages need not be in color; pages that include Websites, driving directions, email, recipes, everyday documents and so many others. Even students' reports usually do not need color, though color sure looks nice. Sure, there may be occasional preferences for color when printing charts and graphs, as well as color transfers for t-shirts, but these, too, are mostly a rarity for most users.

Even if using just the black ink in a color printer, the black cartridge still costs from as low as $4 for generic ink to as much as $25 for printer manufacturer-branded ink, and the yield is so low, at just 200 to, perhaps, 450 pages, with a few yielding as many as 600 pages per ink cartridge. I know there are some printers that will not print if just the black cartridge is installed, or if one or more cartridges of the other colors are low. And I know there are some printers that use individual color ink tanks and some that use one for black and one for all color.

Also, it is my experience that when an inkjet printer is little used, the print heads can become clogged, requiring one or more cleaning cycles, which, unfortunately, becomes costly due to the amount of ink used during the printer's cleaning cycle. There is just no easy way around this and no way to predict when this might be *automatically* required. This is yet another reason to think about whether color is *needed* at all, even for printing the occasional photo.

Contrast printing very few color or black ink pages due to cost and inconvenience versus printing with abandon when it is so dirt cheap to print using a black & white laser printer for day-to-day printing needs. Because it is really so very inexpensive to use *and* buy a black and white laser printer, users can print more pages, not even thinking about the pennies it costs per printed page.

Color photos and graphics *can* be printed in black and white on inkjet printers, though they look less appealing without color. For personal use, or even for reports, that plain look as printed on a black and white laser printer may be just fine, especially when cost is considered.

My advice is this: Get an inexpensive, high performance black and white (also called monochrome) laser printer, sharing it with the family and co-workers within the same home or workplace network. If color printing is a persistent need, but still low volume, consider an inexpensive color inkjet printer, connecting and using it when needed. Better yet, get an inexpensive one of each, both wireless networkable so both are always available to a single user and to a few more. This is easier than you might think. I will only buy wireless printers for use here at *Gadget Central.* Everyone can choose whichever printer is needed for each printing job.

I use and recommend Brother black and white laser printers above all others. They have proven to be of outstanding, long-lasting quality, and delivering excellent print quality with very low cost to both purchase and maintain the printers. One current example is in their **HL-series models**, often found with special sale prices for as little as $70. This, for a printer that can print as fast as 24 pages per minute! Supplies, mostly just toner, are so very inexpensive, in the vicinity of as low as $45 for a 2,600-page capacity NEW Brother-brand toner cartridge, and only about or less than $20 for a generic, compatible toner cartridge designed to be used with that printer. Other Brother HL-series models feature an available 8,000-page toner cartridge, a generic, compatible version of which is available for about $20!

Not only is the cost per page in the vicinity of 1¢ each, but think about the convenience factor. Users can print about 16 reams of paper (500 sheets per ream) before the printer needs attention in the form of fresh toner! Yes, I know the other consumable is a toner drum, at a cost of about $40 for a compatible, non-Brother product lasting about 12,000 pages. For the Brother printers with up to 8,000 page toner cartridges, the drum replacement cycle is about 25,000 pages with the cost of a non-Brother brand, yet compatible toner cartridge costing as little as about $30. It's not a typo!

I have consistently, regularly used remanufactured, compatible cartridges without a problem. I know there have been occasional issues with off-brand toner cartridges. In most cases, the reputable seller will replace the defective cartridge though this can result in inconvenience, of course. What can go wrong? If there is a problem with the new or remanufactured cartridge that is replacing one that is spent, the print quality suffers dramatically within a few pages of installing the new cartridge. Again, this is a rarity, to the best of my knowledge.

Contrast this with printing as many (or as few) as 450 color OR black pages per ink cartridge at a cost from as low as $4 to as much as $25 each. What a huge dollar difference!

For extra savings, some Brother models, including color *and* black and white models, as well as those from other makers, feature built-in duplexing capabilities. This is a setting whereby both sides of a sheet or paper are printed internally, resulting in a big savings on paper.

And don't fear a "refurb" when looking for one of these printers. Factory refurbished models cost less and come with the manufacturer's full warranty as if new. Just be sure of the warranty terms before purchase. Find them online! I know of many users who have purchased Brother refurbs online without a problem of any kind, and the users sure liked the savings!

Inexpensive color inkjet printers are available for as little as $50. If you buy one, use it when *needed*, and either buy a round of ink OR simply buy a new, similar or updated printer when ink gets low, as new ink comes with the new printer, though it is usually in starter cartridges that are NOT full-fills as are their replacements. Another option is to spend as little as about $100 for a higher quality non-disposable printer. Do the math to see the economics of these choices to help in your decision, and include the cost of replacement ink, both from the manufacturer and from third-party suppliers of compatible ink (all purchased online).

If there is a true need (or *want*) for color photos, did you know that better-than-inkjet photo printing is available from most drug store chains, as well as Costco and Wal-Mart, for very little money? Again, the quality is *better than inkjet* and the cost is so much less that can be done at home, unless for occasional single or small multiple jobs. At the rates charged on the outside AND with the better quality, why *ever* print photos at home, except for the occasional single or few photos that may be needed? Sure, take into consideration the convenience of home printing weighed against the possible expense of travel to and from the store where photo prints can be picked up.

Check current pricing online at the printing resource that appears best to you, whether it is in-store pickup or service by mail. Deals offering lower prices appear with some regularity. Just now the Costco price for 4x6 in-store pickup service is 13¢ each up to 499, then just 10¢ each for quantities over 500. Check the specifications for your color printer or the one you may wish to purchase and do the math. It *may* be less expensive and more convenient to do the occasional color photo print at home. For quantity printing, however, I am certain it will be less expensive and less time consuming, and of better quality to print from online services or in a local store, especially with discount coupons offered on many Sites.

One more reminder, though, that an idle inkjet printer is apt to experience print head clogs that can do in the printer or, at the least, be costly to run the automatic cleaning cycles.

Printing through retailers as noted above, did you know that digital photos could be uploaded through a user account and then picked up locally? Sure, you may also bring in a CD, DVD or USB stick loaded with photos and handle everything in the store, too. For even more flexibility and convenience, some outside services accept uploaded that can be set for pickup even across the country, near the recipient's location, all at no added cost. For example, I can take family photos here in SoCal, upload to my Wal-

Mart or Costco account, or any of the others with a real store to visit for pickup. Then, I can specify that I want the photos available for pickup (by friends or relatives) in, say, Florida, New York or any other location where the retailer has a presence.

Prices for larger sizes are also cheap, cheap, cheaper and of the same better, long-lasting quality than can usually be done on a home printer. I can also pick up photo prints locally and send them by regular mail to anyone, anywhere. And remember, the quality is outstanding. Also check prices at **kodakgallery.com, shutterfly.com, pepphoto.com, winkflash.com, riteaid.com, cvs.com, walgreens.com** and any others that can be found. For those that are mail-order service only, be sure and take into consideration delivery costs! Many services offer some quantity of free prints just for signing up, a deal that cannot be beat!

I rarely want to or need to use my color printer for photos, but having an inexpensive color inkjet model has other uses. I recently printed in color onto Avery inkjet transfer paper, which I used for ironing onto white t-shirts. It did a fine job. One last thing on this topic – photo printing at best quality is a slow process on affordable color inkjet printers. Each photo can take a minute or more for each 4x6 photo. Quantity printing can take hours, and even longer for larger sizes in quantity at home. I've experienced color printers taking more than 5 minutes for a single 8x10 at best photo quality, and consuming copious amounts of ink.

What if color _scans_ are needed? Most black and white all-in-one or multifunction printers' scanners are color-capable. I set color docs or photos to be scanned and choose color when needed. It is as simple as that. I set up for scanned docs to be saved wirelessly in a file on the computer setting up that scan, whether a Windows or Mac OS computer.

My recommendation is to start with just one printer – a laser black & white. For students, this is a must. I've followed this plan with each of our three _Gadget Kids_ as they've gone off to college. None have needed a color printer throughout their college experience. Any occasionally needed color printing was done for a modest fee on others' printers or using printing credits that most colleges and universities provide their students, who then use the college's printers. I promise you that many dollars, perhaps more than $300, will be saved with my plan over purchasing and supplying ink for a student's inkjet printer. I can also assure you that my costs for printers and printing was insignificant as compared with friends' who opted to provide color inkjet printers to their children for their school needs. Sticking with laser black and white printers for student use also prevents headaches!

It is also rare in my experience that students who have a color printer get by with just one color inkjet printer throughout the college experience. Something invariably goes wrong necessitating a fresh color printer purchase at least once, sometimes more, during a four to five year undergraduate experience. And it is not unusual for the student to need that inkjet printer replacement in a pinch, without time to shop online. So, it is probably going to cost more from a local reseller, too. I have found that the small-footprint Brother black and white laser printers purchased for each of our children when they left for college have lasted well beyond the full college experience, that is, over four years.

The same guidelines apply for home and small-office use.

All-in-Ones – These are multifunction units that print, scan, copy and may also fax. My recommendations are the same for these products. Unless you really *need* color, get an AIO for home or small office in the form of a black and white laser product, also from Brother. If you *need* color for more than occasional use, get a stand-alone inkjet printer or an AIO (multifunction) unit with duplex, ADF (Auto Document Feeder), wireless and other useful features that may cost only about $100 - $150, making it a worthwhile purchase. Just make sure to get that one laser black and white as the workhorse first, then add a color model as needed. Both can be available to multiple users to select either the black and white laser or color printer each time printing is needed.

Where high-volume color printing is needed (except color photos), check prices for color laser printers *with* the cost of supplies before making a decision. I still like Brother for color laser workhorse printers, but please, do your own math and compare other brands.

Here at *Gadget Central*, our main AIO is a **Brother MFC-8890DW**, with a list price of $500, but available at great savings from online merchants, including refurbs. This black and white laser model does it all, even performing automatic two-sided scans, both black and white and in color, of course! Our MFC-8890DW is used by everyone, both wired and wirelessly, with all the built-in capabilities. Those needing less robust performance should look at Brother's other black and white AIO models, with suggested retail prices as low as $250, so you know they are much less with careful online shopping. These models are often found for less than $200.

We also have a **Brother MFC-J825DW** Inkjet All-in-One, available for as little as $100 up to its suggested price of $150! In addition to its wireless capabilities for all users copying and color printing, scanning, and faxing, this gem also has the distinction of being among few such color printers on the market with the built-in capability of printing directly onto printable CDs and DVDs. Yes! We can create and print beautiful "labels" on these white-topped printable discs. (NEVER use stick-on labels as they are likely going to warp or peel up, which can ruin the drive into which that paper-labeled disc is inserted.) From Brother, only this model and the **MFC-J835DW** have the on-disc printing feature. The MFC-J825DW also has a WebConnect Touch-Screen for direct access to a user's Facebook, Flickr, Picasa and Google Docs accounts. It also can be used to *wirelessly* print from iPhone, Android and Windows Phone 7 mobile devices using Brother's FREE iPrint&Scan software. But wait, there's more! Also included is a USB connection and media slot for directly accessing and printing photos from a digital camera or from most popular camera memory types.

What does this model lack? There is no straight-through paper path or paper bypass allowing one or more sheets or an envelope to be printed without having to load this occasional-use paper or envelope outside of the paper tray. All paper must be loaded into the paper tray. And it is not the fastest at printing photo-quality documents. I can easily forgive these because of the $100 to $150 cost. Replacement ink is also relatively inexpensive (for a color inkjet printer), and there are available 600-page cartridges at not much greater cost than the standard 300-page ink cartridges. As there is no straight-through paper path with a rear exit for the paper, this means that all paper and photos run through the normal path, ending with a 360° route from the paper tray to the exit area, above the paper tray. I would prefer if photos were delivered through a straight path from the paper tray, through the printing mechanism and then straight out the back, without bending the paper. On the other hand, I have not noticed a major curl in any printed paper. Still, this would be my preference.

All in all, I appreciate the features and benefits of this inexpensive color AIO, which anyone here may use wirelessly after installing the software for Mac or Windows computers contained on the included CD. Oh, and wireless setup was exceedingly easy, right from the touchscreen display. I do not use the fax capability as fax sending is through our big **MFC-8890DW**.

Now that I have expressed my unbridled love for Brother printers, here is another dose of reality. I recognize that the brand is not alone in the field of AIO models and I would not be doing my job without mentioning THIS competitive brand you also might like. **Kodak** has come a long way since they introduced a printer product for consumers just a few years ago.

Their **HERO** series (**bit.ly/wgM1wT** – link shortened for your convenience) offers formidable competition. Kodak's take on color photo printing is borne of their heritage in the photo business. I like the photo output very much on this series. And the color document printing is going to meet with positive comments from buyers and users, as well. ALL are wireless capable, even the base model at $100. Built-in duplex printing is included on models starting at only $130 list price.

All Kodak HERO models print, copy and scan. Adding fax capability starts at $200. EACH model carries *this* added feature – the unit's own email address. This capability allows road warriors to send any document or photo to the printer from anywhere in the world and have it printed back at base. How useful is that! And because the printer recognizes if it is a jpeg (photo) or .doc or similar non-photo being sent for remote printing, the printer automatically detects and sets the appropriate paper source – either paper in the regular paper tray for docs or 4x6 photo paper, for example, loaded in the photo paper area. No one has to assist. The feature is not infallible, but I've seen it work and find no major stumbling blocks toward success in most instances.

Off the general topic but relevant to this discussion, all our fax *receiving* is to a PhonePower VoIP voice line, which automatically receives faxes, saves them as TIFF images and then sends email with these attached files to our designated email addresses. It's very simple and most convenient. Details are in the chapter about **Home Phone Service**.

CHAPTER 5

BACKUP

COMPUTER BACKUP AND THEFT PROTECTION SOLUTIONS TO PREVENT ULCERS

If the extent of your computer dabbling is casual online browsing and other mundane activity, with nothing special created or saved locally within the computer, you might be *semi-safe* not backing up. If something were to happen to your computer and nothing could be salvaged, no data, no settings, nothing personal remained, and you wouldn't care, backing up may not be important. *You will want to skip this chapter!*

On the other hand, if your accumulated work product, photos, music, videos, email and other personal, not easily or downright irreplaceable things exist only on your computer, take heed, *please*! This could well be the most important computer-related information you have or will ever encounter, but not because I've written it. I am not so self-absorbed! It is because I hope to have woven a compelling story that resonates with you. It is not expected that an individual would implement all the multiple levels of protection, but please do what you can to protect your data. I don't want anyone to have regrets for not protecting against data loss.

I promise that you will not enjoy reading this chapter. It is dry and boring, with lots of tech talk. I ought to know – I wrote it! But, I

assure you of the importance of the lessons to be learned for those with irreplaceable data on a personal computer.

Backing up means never having to say things you'd rather not. Backing up means you won't lose the "you" that was created *by you* within your computer.

Backing up means you won't lose your *stuff* that is on the computer in the event of disaster. If your computer is lost or stolen, or *when* (not *if*) the hard drive fails, you will be protected and can retrieve your data, but ONLY if you've *properly* backed up.

ALL hard drives will eventually fail. That's a fact, not a scare tactic. If you've never experienced such calamity, consider yourself lucky, plain and simple. Let me say it again: All hard drives, even the new solid-state drives (SSD) will fail. We all hope this never happens to *us*, of course. If you are among those who have had the experience, you know the awful feeling, the helpless, hapless, devastated feeling. Still, even having had the experience, have you changed your behavior so as to avoid this from happening again? Have you implemented a plan to protect against data loss? I certainly hope so.

If you've had a failure or loss, even with backup, it is nearly impossible to avoid some feeling of dread. If you do have a backup, the ill feeling subsides only after everything is restored and returned to its proper place.

Maybe you've heard the term *backup*, but you're not certain what it means. Please allow me to enlighten you with information others typically do not!

Backing up *may* mean copying your personal info, the documents and other "work product" you've created, along with the photos, music, videos and all the rest, *except* applications and *except* the operating system, with all its updates and personalized settings, onto a hard drive different from the *main* drive in your computer. This partial backup scheme is not the best choice, as I will explain, though others would suggest you are well protected. Surprise – YOU'RE NOT!

This could be most easily accomplished with a hard drive next to your computer, attached by a cable. There are other ways to do this, all more technically challenging, yet popular with the more tech savvy (and that will not be detailed here).

If you use a desktop computer that has accommodations inside for additional hard drives, such an added internal hard drive could be your "local" backup drive.

Backing up could also mean using an outside service, far away from home, to which your *stuff* is backed up over the Internet.

Backing up could mean both local and far away schemes, which is the best way – local *and* "in the cloud," also called *online backup*.

Let's back up just a bit, however! **If you don't back up your computer data and disaster strikes, what can you do?** The hard drive has failed, and your computer will not start up. This can and usually does happen spontaneously, without apparent warning. From experience, I can tell you that this causes a deep sense of immediate loss bordering on panic. It is NOT funny! All your stuff, poof, gone! And let us not forget that data loss can also be the result of the theft of a computer or the result of accidental deletion of one or more files.

Here's the saddest tale of all. Even if you've done your part, backing up locally, a fire, flood, earthquake or tornado can wipe out everything in the home, original and backup, at the same time. Let's not go there or we'll both go crazy! Stay calm and proceed, but know this advice is going to benefit you even in the worst-case scenario.

Without backup, you are left to recovery of the data on the affected drive in the first and last examples above – spontaneous drive failure and accidental erasure/deletion of one of more files. There are numerous software programs claiming to be able to retrieve erased data from a working drive or data from a failed hard drive. Sometimes hard drives experience intermittent failures before completely going south. In short, however, and in my experience, no software program can *assure* the ability to fully restore data from a drive that has failed as well as one with data that has been mistakenly deleted. I've tried, spending weeks experimenting and hoping, using data retrieval software purporting to be able to do the job. After all my efforts with such software, I was unable to recover the data from the on-the-fritz drive. It was a frustrating experience, one I will not go through again. Life is too short to spin one's wheels in such a likely fruitless effort, on a maybe. In the case of a small amount of accidentally erased data, it might be possible to recover if the deletion is caught soon after it occurs, but this, too, is not assured.

There is one virtually certain method of retrieving data from a gone-bad hard drive or from one on which data has been inadvertently erased. **The ONLY way to do this that I know and trust is with services from a company in the business of data recovery. The ONLY one I use and recommend is called *DriveSavers Data Recovery* at drivesavers.com.** Leave it to the experts! Located just north of San Francisco in Novato, CA, the company has received many honors in their approximately 30 years in business and is trusted by entities from government agencies to top companies worldwide, as well as thankful, grateful individuals including your humble writer. *Call* **them at 1-800-440-1904.** These are the experts I have trusted and used for more than 20 years, whether for Macs or Windows PCs. They are THE experts, as well, who can retrieve lost data from any other form of computer-type media – camera and phone memory cards, iPods and other MP3 players and so on. This is the good news.

Yes, it is likely that DriveSavers experts can recover your precious lost data from a hard drive. The downside is that it is expensive. On the other hand, they get results. Company co-founder, Scott Gaidano, and I have spoken many times about data recovery (he's also a fellow flashlight-aholic – see the chapter on **flashlights** – and one heck of a great guy). I've heard and read stories of amazing data recovery accomplishments in the company labs, some of which are in testimonials on the company Website. **The cost to recover data from a hard drive can easily top $1,000, and possibly $2,000.** You'll need to discuss with them the possible cost for partial recovery. Still, it is not going to be inexpensive. I'm sure this gets your attention. **Prevention is what this chapter is all about.** Scott agrees and urges everyone to take necessary precautions so we don't need his company's expert services.

Remember, the experts at DriveSavers are there in the event of need, either for personal or business computers or memory devices. Wouldn't you rather be safe and protected against this need, to the best capability you can muster? **Even if data is recovered by DriveSavers experts (or by any other recovery service),** it is still up to the user to discern from the totality of recovered data exactly what is important and what is not and then to redistribute it in its proper place. This can still take copious amounts of time in addition to all the expense. Point made, I hope!

Before alerting you to reasonable-cost solutions, think about what *you would do* **with a backup in the case of needing to use it?** <u>Now we are getting to the crux of the topic.</u> It is one thing to advise that backup must be done and quite another to explain in detail what must be done to make your computer "whole" again from the backup. This is the part about which nearly everyone who has not gone through the data restoration process has given little thought. The simplest case is if you've backed up either continuously or regularly and the hard drive fails inside your computer. This could also be if something is inadvertently erased from the object drive and not yet from the backup.

If a drive fails, replace the failed drive with one readily available from local or online merchants. If under warranty, you're covered, but only for the labor and failed part as close to the original as possible, and not usually for an upgraded hard drive and *NOT for the recovery of data on the failed drive*. **This could be a HUGE problem!**

The capacity of the failed drive is imprinted on the drive's label. You may also wish to install a higher capacity drive, likely a relatively inexpensive non-warranty upgrade. Recommended brands include **Western Digital (wdc.com)** and **Seagate (seagate.com)** as my go-to-choices. If you are not sure of a compatible replacement drive choice, you may wish to contact the companies by phone for a recommendation. Provide the make and model of the computer in question, so have that info handy before contacting the companies. OR, as a precaution, get the info and make the call as if the need is NOW, then keep it handy on paper and in a location you will see if disaster strikes. (Paper? Yes, paper. If you store the info digitally on a drive that fails, how will you retrieve it? Paper is the answer!) Also, remember, the info you find today is probably going to change in a year, two or three, or more, as these companies update, improve and upgrade their products, providing higher capacity at lower prices. Once the replacement drive type and size is determined, use the product number as the search term to shop for it online at the best price if you have the luxury of time.

On *most* modern computers, replacement is an easy do-it-yourself job, taking just a few minutes, assuming you have the proper size screwdriver and/or other appropriate tools. Go online and look up the replacement procedure of the hard drive in your current computer for details to help in making the decision to do it yourself or not. The information is probably a few clicks away. Alternatively, ask a geek friend or, in the worst case, inquire of a local professional shop or Apple Store in the case of a Mac for the cost to purchase and have a hard drive replaced. Compare cost estimates for Mac hard drive replacement between the Apple Store and local non-company Apple specialists. The same kind of drive is in both Windows PCs and Macs, so you need not be concerned with a Windows-only or Mac-only internal drive. It is how the drive is formatted after installation that determines whether it is for Mac or Windows use.

Once the drive is replaced and *if* you have startup and recovery discs for your computer, you would start the computer from the startup disc. And IF you've done your backup locally as described below, you might be good to go by using that backup as the source from which to repopulate the new hard drive. It *could* be that simple.

Recall that I do not recommend the partial backup routinely recommended and promoted by others. Without a *full backup* of the type about to be recommended, users are in for an unpleasant and time-consuming journey. Partial backup is NOT very helpful, especially if faced with a disaster that necessitates using it.

Macs are a computer breed unto themselves. Recovering from a Mac hard drive failure or accidental data erasure is simple and straightforward, so long as Apple's built in backup capability called Time Machine is in use. Explanation coming up!

Though NEW Macs, with the introduction of Mac OS X 10.7x Lion, are no longer provided with discs, Apple's buyers are covered, so I will not go into details here. Users of these new Macs can download whatever is needed to get the process started. Macs purchased when recovery discs were included will also be easy to recover. Apple Store personnel and Apple phone support can instruct in how to create a startup disc, if needed. *My advice is ALWAYS to be prepared ahead of need!*

Most Windows PCs do not come with everything needed. No included startup disc? For *your* Windows computer, look up and look into how you can create one or otherwise acquire one. Ditto the recovery discs. Now would be a great time to become prepared for such need by communicating with your computer's manufacturer to learn how to get everything needed to set up your computer with a new hard drive and to restore it to factory-new. It might be easy or you might have to jump through hoops to get what is needed. With so many brands and models of Windows PCs, you are somewhat on your own to do the research and disc procurement, but DO NOT fail to complete this task, please, in advance of need. And the next time you buy a new computer, be sure to take care of this at the time of or shortly after purchase.

Regardless of whether you have a Windows PC or a Mac, if it is avoidable, I don't want anyone to have to build-up a computer from scratch, that is, to start fresh. This is not only very time consuming, it is also a maddening experience to be dreaded. There are some circumstances that are best dealt with by doing a nearly ground-up re-do of a computer, however.

For example, if you've a Windows PC that is hit with a virus, I assure you that the only way to most certainly eradicate it is to reformat the hard drive, and then to install the operating system, then all the OS updates. Here is another heart-stopping realization. For all your efforts to carefully back up Windows PC data, if it is a virus and not a hard drive failure that is the reason you need to start fresh with the existing internal drive, your backup is most likely also contaminated, so don't simply reformat and repopulate that drive from the backup. This WILL NOT FIX A VIRUS PROBLEM. Restoring a Windows PC from a backup point that is pre-infection is a great idea, but knowing what is that point may be problematic. I have no ironclad solution.

If a Windows virus is the source of your trouble, then it is likely you have not been using a good antivirus program OR what you have been using has not been automatically updated daily, even several times per day. When you have finished reformatting, then freshly reinstalling all your original applications from scratch using the recovery discs I said would be needed, after hours and hours, be sure to install the antivirus program of choice or per my recommendation, my preferences to be revealed later. This is a good time to do some house cleaning on that computer, by not installing apps you no longer want or need, and getting new versions of those you do want and need. And then download and install all the other apps determined to be essential, or install from the original discs, if available, then be sure each is updated before installing the next, and so on. It's a ton of work, usually taking a day or two, or more.

Next, attach the backup drive and set the antivirus program to scan the backup drive's folders – Documents, Music, Video, and so on AND to repair any detected viruses before dragging the contents of those folders to your new same-name folders on the reformatted drive. It's a time-consuming and labor-intensive process I hope Windows PC users never need to do! You will not be pleasant company to anyone during the process. At the very least, I hope you will be incentivized to use antivirus software that automatically updates as needed. And since viruses can rely on users' behavior that allows entry into a computer, I hope you are more careful in your computer use. NEVER open an attachment from an unknown sender, for example. And NEVER click a link from anyone you are not certain is a scammer.

Unfortunately, some viruses are sent after a computer is compromised. The user's contact file is often co-opted, appearing to send what seems to be from a friend or colleague to you. This fake email can contain the virus bomb that brings down your system and many others'. And I haven't even mentioned the other obvious point to many of you, that all your old email on the contaminated backup drive (unless all your email is online, such as with Gmail or Yahoo mail) is also suspect and may be infected, and MUST be scanned before bringing it back to the renewed drive. It's just not pretty and it's not fair! I'll bet you really, really, *really* want to avoid all of this. As there are currently no viruses affecting Macs, this is a procedure not likely to be needed on Macs . . . YET.

Another common Windows PC need is to reinstall the operating system when nothing more than a significant slowdown occurs. You'll know when things have slowed dramatically, I can assure you. In this case, reinstalling Windows over the current installation can renew and refresh speed dramatically. You'll need your original Microsoft Windows install disk or the one you have procured from your computer's maker if it was not included when the computer was purchased. The procedure is not lengthy, but I will leave it to you to research the how-to on this procedure.

Even Macs may need occasional re-installation of the operating system, usually an easy operation on a Mac, and without reformatting the hard drive. If the computer runs slowly or if there is some unexpected and unpleasant activity I cannot begin to describe here, the first attempted remedy is always to run the **Fix Permissions** capability in the **Apple Disk Utility** program as described below.

If reinstalling the Mac OS is in the cards, ask about how to do this for your Mac at an Apple Store, look online under the support tab at support.apple.com or find support options at apple.com/contact.

First, and maybe now, see what version of the Mac OS is installed on your computer. Click the Apple symbol at the top left of the screen and drop down on **About This Mac**, the first in the list. In the new little window, just under the BIG and BOLD **Mac OS X** will be written the version of your software. This is the info you will need to properly restore it to this version again. If you have not upgraded the Mac OS from what was on it when received, then all the Software Updates done to now have updated to the version shown. Once the OS is reinstalled, then doing all the updates through Apple's Software Update utility will bring you back to current, to the version you have just discovered. The faster your Internet speed, the better! Again, Apple will help educate users as needed.

Here are a few related tips– clicking the **Version** in the **About This Mac** box will cycle through other useful info (on modern Mac OS X versions); the Build number of the OS and then the **Mac's serial number**, so you don't have to look on the computer's exterior for this info. If calling Apple tech support for help, you may be asked for the computer's serial number, which you now have.

Now, let's **Fix Permissions** on your Mac! **Navigate to the Utilities folder inside the Applications folder**. Double-click on **Disk Utility**. In that utility at the left of the window, you will see the highest level of the main, internal hard drive, one level above **Macintosh HD**. Select that highest level of the internal hard drive and, at the right, near the top, click the tab that reads **First Aid**. Near the bottom of the big window you will see both **Repair Disk** and **Repair Disk Permissions**. Repair Disk will not be available. As a maintenance item, select the **Repair Disk Permissions** choice.

Disk Permissions repair may be done through the Disk Utility at any time after a normal startup, and is recommended (by me) to be done with some regularity, perhaps once per month. Restart your Mac just to see if it is a bit speedier than when you noticed it was running more slowly. If it is still slow, seek advice at the Apple Store.

And now, back to the disaster recovery . . .

As we continue, it is important to present a worst-case scenario to give you such an unpleasant taste for it that you will take necessary steps to avoid having to perform these steps. To do so, we need to explore how to make use of an incomplete or partial backup, that is, backed-up data only, putting it back where it belongs on a new hard drive, or, perhaps, on a new computer. Most computers are set up with folders into which documents are stored. Most computers are automatically set to put applications, music, photos and videos in their proper place, as well. The backup is going to be similarly configured. If the backup is a locally connected hard drive, one that uses USB or FireWire to connect to the computer, the *contents* of the many folders on the backup can be dragged to the associated folder on the newly minted replacement drive. It can be that "simple." Depending upon how much data there is to be transferred, this can take from minutes to *several hours*. You don't really want to do all of this, do you? Help is coming soon!

Some backup schemes include simplified restoration instructions that automate the process, whereby the software knows what to do and does it. I'll leave you with this before venturing into better, simpler backup methods: For your own wellbeing, become familiar with what it is you will need to do and how to do it *before* you must do any of this. **Be prepared so you won't get an ulcer! There is more to come on this, so read on,** please.

How can you back up your computer? I'm so glad you asked!

Mac users have many options. The one provided for ALL modern Macs is built-in, an application called **Time Machine. Get an external hard drive of at least the same if not more capacity than the internal drive, the one to be backed up.** How can the drive capacity be determined? On Macs, click once to highlight the **Macintosh HD** on the desktop. Then, click **Command+I** to **Get Info.** The approximate capacity of the drive will be in the long window that opens – see the number next to **Capacity.** My advice is to get a drive for the **Time Machine** backup purpose that is the same as or one notch greater than the capacity of the internal drive.

On Windows PCs, go to **Control Panel**, then **Device Manager**, then click to turn down next to **Disk Drives.** Now right click on the HDD, usually with a "0" next to it and select **Properties.** In the new window that opens, click the tab called **Volumes.** Click **Populate** near the bottom of the window and then look above and at the left for the word **Capacity.**

THAT is your number, though usually listed in MB, which can be very confusing. The computer's hard drive is everywhere else called out in GB or gigabytes, each one of which is 1,000 MB, or megabytes. You've just got to love Windows! Here is the easy translation toward GB from MB – lop off the last three digits. For example, if the number is in the vicinity of 496404 in MB, that would be about a 496GB drive, close enough to 500, which is likely the size of that drive, as there are no 496 GB drives. Use this as a guide, but always round UP, not down, and you are certain to find the capacity number needed.

Regardless of the computer, Mac or Windows PC, the same kind of external drive may be used for backup. The drive need not be one specified as a Mac-formatted drive to be used with Macs. The cost can be well under $100 to about $200, depending upon its capacity. Learn what are the connections on your Mac. If for use on a Mac, be sure the external has a connection compatible with the Mac, meaning USB 2.0, or, perhaps, FireWire

800 or, on the newest Macs, Thunderbolt. Thunderbolt is quite new and it is difficult at present to find inexpensive and plentiful sources of drives with this new and super-fast connection. External drives marketed specifically for Macs usually come with FireWire connectivity, plus USB 2.0. If you are a Windows PC person, and if you have a 2011 or newer PC, it is possible that yours is equipped with the newer USB 3.0 technology. If this is the case, check prices for USB 3.0-equipped external drives. If the price is right, go ahead and get USB 3.0. If that cost is too steep, get a drive equipped with the more common USB 2.0. All Windows PCs and Macs are compatible with USB 2.0.

My advice is that the external backup drive is to be used exclusively for backups. The backup will be used to restore after an internal drive failure or to recover files inadvertently removed from the internal drive, so the speed of the backup drive is not going to matter much. The operation is still going to take from an hour to three or more for a complete restore, depending upon the size and the used-up capacity of the internal drive. In other words, just because the drive's capacity is, for example 1 TB, if only 240 GB has been used, it is not going to take as long as if that drive had 750 GB of data on it. If using a USB 2.0 drive on either platform, the first backup can take as much as 12 hours, even more, if your drive is nearing its useful capacity.

Time for another USEFUL TIP!   Never, *never* use a hard drive to within less than 10% of its total capacity. Danger is ahead! If a 500 GB drive, for example, has MORE than or close to 450 GB used, it is time to do one of two things. Either dump unnecessary apps and/or files and documents, (cleaning out a huge, onboard email box may free up significant space) OR get a new and larger capacity internal hard drive. Even as you get close to the 10% free space number, it is time to do some serious house cleaning. Photos, a large music library and video files are the biggest space hogs, but you may want all your photos, music and videos, and why not? If this is the case, you may wish to get a separate drive onto which those old and rarely used video and photo files may be stored. OR, you may want to spend a weekend copying those files to DVDs, carefully labeling them with an **acid free permanent marker or creatively printing directly upon a printable CD or DVD**. Most art supply stores carry acid free markers. ONLY use this type of marker to write on a CD or DVD, and NEVER use a paper label on optical media as it can dislodge and ruin the drive into which the disc has been inserted. Using a non-acid free marker might ruin the disc by the acid penetrating the layer below the top of the disc. Just don't do it on any disc you want to last as long as possible. It is not worth the chance for failure.

Use printable discs if you have a color printer featuring direct printing onto these discs. Brother currently offers printers with this built-in capability.

Before use, format the external drive. To format a new external drive on Macs, plug it in to power (if needed) and connect it to the computer. Find and launch the **Disk Utility** application in the Utilities folder. Find the new drive listed on the left. Click to highlight it, certain it is not the one listed as Macintosh HD, which is the computer's internal drive. You will want to click on the highest level of the new drive. Now, at the right, see the few tabs at the top of the open window. Click the one named **Partition**. If that tab is NOT there, you are on the lower, indented indication of the new hard drive. Move up a level. Once Partition is available and selected, there will be a drop down menu showing **Volume Scheme**. Select 1 Partition, then **Options** near the bottom of that window. In the **Options** window, select **GUID Partition Table** if the Mac is running Mac OS X 10.5 Leopard or later on an Intel-based Mac. Older PPC-based Macs should use **Apple Partition Map**. This will format the drive to allow it to be a startup drive for your Mac. Name the drive in the space provided as, perhaps, **Time Machine Backup**, so that drive's purpose will be obvious. Click **Apply** and in mere moments, the drive will be ready. **Quit** the Disk Utility application after formatting is completed.

Find the **Time Machine** application in the Applications folder (if the Mac does not automatically recognize that a new and properly formatted drive is connected and asks if you want it to be used for **Time Machine**). Double-click! That app will display a panel allowing you to turn on **Time Machine**, so do it. Now, be sure that new **Time Machine** drive is the selected disk and set the options to *exclude nothing*. Then, let 'er rip. Also be sure to click the box to **Show Time Machine status in the menu bar**. This continuous Time Machine backup takes *snapshots* every so often and maintains data backup versions as you just do your thing. Confused? Find help from the same recommended Mac resources above.

In the event of drive failure, the internal drive will require replacement. If you've used Time Machine regularly, you should be in good shape with little or nothing lost. Proceed if comfortable or ask for help. If your Mac came with discs, use the first one to start it up and double-click to install the Mac OS X icon. In the Installer, choose **Utilities > Restore System from Backup**. In the **Restore Your System** dialog box, click **Continue**. Select your **Time Machine** backup volume. Select the most recent Time Machine backup. Then follow the onscreen instructions. Sit back and watch and wait as everything, including the operating system and all system data, as well as the applications are restored to the most recent state. No updates to do, no personalization settings need be performed. It's all there in Time Machine. Mac users please learn more by searching online for *mac 101: time machine* to find the informative Apple support document on this topic. It's fascinating! Complete instructions are there. The first complete Time Machine backup may take several hours, so you might want to start it before bed and check in the next morning. Be sure to NOT exclude anything during or after setup. You'll want a **complete backup**. And be sure to read about how to use a Time Machine backup to restore to a new or newly reformatted hard drive. You will see the exact instructions listed above, as they are taken directly from Apple's online info.

New Macs running Mac OS X 10.7x (Lion) or those who have upgraded a Mac to Lion might not have a disc, as none was included or available, unless one was made (following instructions found by searching online for **creating lion install disc)**. Apple expects users to have downloaded the Lion install disc and to have burned it onto a DVD as found using the search criteria in bold type above. It is also possible to create an Apple startup on a USB thumb drive or to buy one already configured. Still confused? Visit an Apple Store or, if still under the warranty period from the new Mac purchase or the purchase of Lion, call Apple for tech support and get instructions.

The downside to this or any continuous local backup plan is that the external drive must be connected either always or at least with great regularity to be current. Remember, the completeness and currency of backup is only as good as your most recently completed backup. If your drive fails after your last backup and before your next one, all that has been created and saved since the last backup will be lost. You wouldn't want the failure to occur after the last backup and before the next one, would you? One possible workaround is to connect the Time Machine Drive and click the dropdown Time Machine menu at the top of the screen where it says **Back Up Now** after any project completion. If content is created on a laptop while away from home base, and if what is created is not too large for your email, send it to yourself. Once in email, whether Gmail, Yahoo or some other service, the file or document will be stored safely online, giving you a file backup window until your actual Time Machine drive is connected once again. The Time Machine drive will recognize what's new and back it up. Then, you can delete the email containing that now redundant file. Windows PC users are encouraged to also email files created while away from base.

If Internet access is not available while away from base and creating content, whether on a Mac or Windows PC, purchase a USB thumb drive of sufficient capacity to accommodate any planned need. Then, save the file to that drive. Just be careful to protect the thumb drive. I have always had excellent results with the reasonably priced **Kingston USB thumb drives** (**kingston.com** – click on **FLASH CARDS & DRIVES**, then click to see the selections, and search online by the Kingston part number for the capacity of interest, such as DT101G2/16GBZ for their 16GB DataTraveler101). Other reliable brands include **SanDisk**, **Lexar**, **Corsair** and **Verbatim** among a sea of other makers. I use a 32 GB Kingston USB thumb drive so there is plenty of space for almost any purpose. **I also suggest a cap-less design so there is no cap to misplace.**

Do USB thumb drives last forever? I've never experienced a failed USB thumb drive, but I know it can happen. My advice is to not rely on long-term storage using one of these devices. DO use them for the purposes detailed here. Just remember that they *can* fail. They are also incredibly rugged, with many stories told about such drives successfully making it through the washing machine and dryer unscathed and very clean.

Though Microsoft includes their own **Backup** capability in some versions of Windows XP and for all users of Vista and Windows 7, their capability does not provide a complete backup, including applications and system files. Pass and move on. I do not recommend using it!

The same external hard drive size rule applies for Windows computers. The purchased drive will likely already be formatted for use with Windows computers, but I would still freshly format it. Do this by connecting and powering up the new external drive. Then, navigate to My Computer or Computer (in Windows 7). Select **Open**. The new drive *should* be included in the window that opens. Right click on the new drive and select **Format** from the menu. In the new window you will see formatting choices, including a default selection if in Windows 7. Also, you can create your own **Volume label** (the name of the drive). How about **Backup**? Choose **Quick Format**, click **Start** and you're done.

Mac users have **Time Machine.** For Windows users I recommend **Rebit 5, (rebit.com)**, the closest thing to the ease and simplicity as well as completeness of **Time Machine**, but exclusively for Windows users. Search for the best online price for **Rebit 5**, for one PC or for three PCs. The online shopping price I just found is about $25 for one PC and about $60 for the three-PC version. It is also available pre-installed on either a 1 TB or 2 TB drive, but I suggest buying a drive separately as a cost saver and as it offers greater flexibility of drive capacities. Note that the pre-installed drives are set to accept up to three backups, which I do not like. I want one backup drive dedicated to ONLY ONE computer. Installing and using **Rebit 5** on a new, newly formatted Windows drive is as easy as it gets, almost as easy as setting up Time Machine on a Mac. Install, follow the prompts for a **complete backup**, with **nothing omitted** and let it start working.

Remember that regardless of whether on a PC or Mac, that first backup is going to take a long time, perhaps several hours, so just let it happen. Preferably, start the process at night after you are finished playing or working on the computer for the day. Have a restful night and more thereafter, secure in the knowledge that you are on the road to lower-stress computer disaster recovery.

The automated backup and then recovery process, other than being time consuming by necessity, works well for Macs (**Time Machine**) and Windows PCs (**Rebit 5**). Once plugged in, they just work!
To review, these backups for both Mac and Windows PCs allow nearly idiot-proof full and complete backup and recovery from disaster due to internal hard drive failure. This is also a great way to backup and then restore to a larger internal hard drive than was the original. Out with the old drive, in with the new!

Use **Rebit 5** to create a bootable Windows recovery USB stick, start the PC from it, then start the simple process of setting up the new drive from the Rebit 5 backup, which, when finished, is up-to-date and complete as was the original in all regards. For Macs, start from either a restore disc or a startup disc as instructed by Apple for users of the newest Mac OS X Lion 10.7x, then restore to the new drive from the Time Machine backup. **If used when making the move to a larger internal hard drive, be sure to get a commensurately larger new external backup drive!**

So much for traditional, *local* backup. I also want to bring to your attention non-traditional backup available for Mac users only, one of which I really, *really* like, but it is somewhat controversial, at least among some folks I know, in that they specifically recommend against it. Why? Some feel it can create a false sense of wellbeing, plus, it demands strict adherence to at least ONE rule: The backup drive may NEVER be used for anything other than this singular purpose! Period. Once the drive is used for any purpose other than what will be explained shortly, it is no longer a safe and protected backup drive. Despite this caveat, this same rule applies to ANY drive for ANY computer that is used for backup. That's a big, "Well, duh!" Never use your backup drive for any other purpose. I like and use the following plan, but only *in addition to* Time Machine on my Macs. I like the idea of redundancy and am comforted by it.

What if you could create a *bootable* backup of the entire Mac hard drive that is an exact copy of the internal drive? By exact, I mean, nearly bit-for-bit, byte-for-byte, an exact *clone* of the internal drive. **Time Machine** backup is not bootable as a Mac drive, though it can completely restore a reformatted existing drive, a new one or it can be used to recover files recently deleted from the main drive. This includes the Mac OS, all data, settings and all applications.

Earlier, I detailed the steps to using a backup for disaster recovery. Now is the time to process all that information, step by step. Once a drive is replaced, it must be formatted and the operating system installed, Windows or Mac. Then, it must be revealed, there is time-consuming work ahead to restore the computer to its state just before the failure. This can take much longer than an hour. However, IF the backup is Time Machine on Macs or Rebit 5 for Windows computers, these drives may be called upon to completely restore the new drive from the back up to the new internal drive, or to a new computer. Everything is back in its place, just as before the loss, while you just wait. Let this sink in, please.

If you thought you were being clever by copying individual folders to a drive you call a backup, you will be quite sorry in the event of a failure of the type discussed here.

In short, everything you did to make that computer yours, every setting, every update, every added application, every little thing will be GONE, unless you do it all again, step by step. That is a dirty little secret of backing up and why I advocate for the complete backup only. Even if you think you've done your part, there is still much to know, much to do and much time needed to do it. The lesson here is that users are best served with the complete solutions that become useful restoration tools rather than partial backups that are not at all useful if a backup is needed to restore a failed drive, upgraded drive or replaced computer.

Back to Macs. Here we go again where we left off. What if ONE backup could be created and maintained that had it all, exactly as your computer was, up to the moment of the last backup? What if that backup was created such that it *mirrored* the "original" drive in every way, changing only what was changed on the original. As such, once the first backup is made, further backups are done more quickly, making only *incremental* changes. In an incremental backup, the main drive is compared with the backup and ONLY items changed on the main are updated on the backup, or clone, though what is erased on the main is also erased on the clone; something not to be taken lightly!

And what *if* the internal drive failed and the computer failed to start up because of it? What *if* you had a plan? Instead of having that sinking, sick feeling of extreme doom, you could connect the "clone" hard drive and start your Mac from that external clone *before replacing the failed drive*. Would this be a time saver? The Mac will start in mere minutes from that external, cloned drive and the computer desktop would look as if nothing had gone wrong! Remember, even if you've done a perfect backup, with all data preserved, it is still going to take a great deal of time to repopulate the new drive from a backup.

On Macs ONLY, users can hold down the **Option** key at the sound of the startup "BONG" sound while rebooting with the external clone powered up and connected. The computer senses the external drive, showing a graphic representation, an *icon* on the desktop, of either the external drive only OR both the external and internal (even if the internal is not capable of starting the Mac). Once the icon or icons appear on the desktop, release the Option key. Clicking on the arrow below the identified external clone starts up the computer from the clone *as if nothing had gone wrong*, though a bit slower than normal.

ALL data, and I mean everything, are right there (on the external drive) just as it was on the internal. Since that is now the ONLY known "good" operational drive, it is critical that only emergency operations are performed and that the misbehaving internal drive be replaced forthwith. Do what work MUST be done and ONLY what MUST be done. There will be no playing! Users are on borrowed time until the drive replacement, which is one reason why critics say this is not a good idea. In this scenario, users may continue to use an attached Time Machine drive. Because it is operating on the clone, Time Machine will think it is operating on the original, internal drive, but **I cannot stress enough the importance of replacing the failed drive ASAP**. Don't moan or whine to me if another failure occurs before the misbehaving drive is replaced.

This is the ONLY way to be up and running in mere moments. The ONLY way. Different from Time Machine, I use this cloning method as an additional safeguard. The only cost is for the additional drive and, in my case, the additional software, though, as you will learn, FREE solutions exist to create this cloned drive. There is nothing else that can be done to allow such instant, though temporary relief from a failed internal drive on a Mac. I know of several users, including in my own family, who have precisely followed this advice and have benefitted from this almost magical temporary fix when the internal drive has, without notice, without warning, failed on startup.

Once the affected drive is replaced, then the computer can again be started from the clone, which can be cloned *back* to the internal after simple formatting. I've simplified the scenario, but there is not much more to it, though these steps must be followed with precision for success to follow. I've done this many, many times, including when I have changed computers or upgraded hard drives – ON MACs only. **Only Macs can start up from an external hard drive! Sorry, Windows.**

There is software available for this purpose, both FREE and for a fee. **SuperDuper**, $30 from **shirt-pocket.com**, and the FREE **Carbon Copy Cloner** from **bombich.com** are both good choices for creating Mac "clone" backups. I use a paid program called **Personal Backup** from **Intego** (**intego.com**), the same company making recommended Mac antivirus software. I like the flexibility and simplicity offered by this software and that I can include steps I consider important that are not part of the free programs' features. In particular, I like to "build" my backup plan from their step-by-step capabilities in the program; First, I select Bootable Backup from the choices on the left, and I remove the other choices. These are the setup steps and choices; Schedule (I choose every other day at a convenient time); Preparation (not used); Exceptions (not used); Bootable Backup Options (I always want to be prompted for an administrator password, so I KNOW the backup is about to run, and I check the box to **Repair permissions** on my destination disk (the backup drive) as well as to repair minor disk errors (this uses the Mac's built-in capabilities for both); and finally, Finishing, where I select to **Unmount the destination** ALWAYS (this ejects the external drive, powering it down) and I select the checkbox to **Quit Personal Backup**, all neat and tidy. **Personal Backup X6** is part of the company's **Internet Security Suite** and not available separately. **Virus Barrier X6**, also recommended, is there, too. I do not use or recommend the other apps in this suite, however, because I feel they are overkill for most savvy users. Why not try Intego's 30-day free trial through links on their Site?

If you like the experience, shop online for the best price for the Intego solution, then uninstall the free trial and install the paid version found at the best price. Or try SuperDuper or the **FREE** Carbon Copy Cloner method. While on the adjacent subject of Mac antivirus software (yes, it's time for Mac users to fall in line with our Windows brethren), I also highly recommend as excellent the new **ESET Cybersecurity for Mac (eset.com)**.

BOTH MAC ANTIVIRUS SOLUTIONS ARE AVAILABLE AS A FREE DOWNLOAD FROM THEIR RESPECTIVE SITES WITH A FREE 30-DAY TRIAL. CHOOSE ONE OR BOTH, DOWNLOAD AND TRY THEM, BUT NOT AT THE SAME TIME, and be certain to use their uninstall capability to remove one before installing the other! The Uninstaller is going to be in the product's package that was downloaded when it was installed, so SAVE THE INSTALLER on these downloads. Instructions are at each product's Website in the support section. Search using keyword UNINSTALL.

And now, a word about FREE Mac antivirus software – **Sophos Anti-Virus for Mac Home Edition (sophos.com**, then click **Products**, then **Free Tools** to find it). Sophos *is* FREE and worth the price. I know users who like it and find it adequate. It is difficult to assess, as there are hardly any malware threats on Macs. I found it to be just OK as compared with *my personal feeling* toward the recommended choices. I *feel* more protected with the others. You may wish to give Sophos a try and if you *feel* you are protected AND if you perceive no computer slowdown when using it, proceed, of course. If you perceive any downside, then uninstall it and try the others. I perceive NO slowdown using any of these, but I have a very speedy Mac. I know users who have reported a perception of tolerable slowdown with Sophos installed on a five-year old MacBook, for example, and I know others who feel totally protected with no perceptible performance hit from their Sophos installation.

On my Macs, I set the **Personal Backup** software to prompt me to back up every other day. That way, I am only a day behind with what's new and it takes less time to perform the backup. I STILL email important files to me in the interim, erasing them once it is confirmed they are on the newly updated backup. When at home base, and at the end of any day in which I have *created* something important, I always do an unscheduled backup before shutting down. I allot about 30 minutes to this task on a 1 TB internal drive that is about 85% full (Yes, I will be upgrading to a larger main drive before too much longer AND finding data to delete, as well).

Because Macs can start from an external drive, this means users should *test the clone*, perhaps after every other backup is done. Start from it using the **Options** key instructions above and be sure all the data is there, all the info in the documents folder, the photos, music and videos, too. I test this by inspecting the "**Get Info**" box on the **Macintosh HD** on the desktop (click it to highlight and then enter Command + I) and doing the same thing with the name for my backup drive, the *clone*. Be sure to use a unique and different name for your clone backup to avoid mistaken identity. Compare both internal and external drives, the original and the clone. Do NOT do another backup until it is confirmed that all the data is on the external clone. Also, while booted from the clone, open several critical apps ONLY FROM THE CLONE and NEVER from the Macintosh HD, the normal internal hard drive, to ensure they are working properly AND that all the expected data is there. Be sure your iTunes music looks fine and is all there, as well as iPhoto. No, you do not have to remember the number of songs or photos. If the programs open and appear to behave normally, they are fine. **If these tests are not done and you are not SURE things are fine, and then if another backup is performed, remember that the external mirrors the internal, so what is missing from the internal will now be missing from the external after backup.**

I made this costly error that is not likely to be in your future, but it's a lesson worth noting! In my case it was when upgrading the Mac OS on a laptop (from Leopard to Snow Leopard). I assumed all was fine on the updated drive and stupidly backed up without checking as indicated above. Only later did I find there was a bug in Apple's software update procedure when I updated the iLife application going from one generation older to the newest iLife '11, which includes the iPhoto application. The iPhoto files were not copied during the upgrade. This was a known and occasional bug in that prior OS version. There is no issue with the new Mac OS X version called Lion.

My only remedy was to elicit the recovery services from DriveSavers. Though the files had been erased from my backup, DriveSavers was able to find and recover *all the photos.* I don't want to go through that again, and I _never_ want you to go though this. The photos, all of them, including thumbnails, were recovered, but as indicated above when discussing DriveSavers and data recovery, in general, they had no way of knowing exactly which photos were ONLY those from iPhoto, so it is still an ongoing process, many months later, to search the DriveSavers recovered data on a separate drive and put the photos back into iPhoto, with all the tags and captions unrecoverable. You *really* don't want to ever have to do this! Thank you to DriveSavers for saving my behind, however.

This technicality, along with the fact that all of this may just be too much for you, is one reason to not proceed lightly with cloning. The first backup can take three or more hours, depending upon the amount of data to be backed up and the size of your hard drive. Subsequent incremental backups, if done with regularity of every other day or more frequently, can take as little as 20 to 30 minutes. This discussion does NOT apply to Time Machine, by the way.

It is a good idea to do NOTHING else while the backup is being created or when it is being updated. If this admonition is not followed, not acted upon, the backup software cannot accurately report errors. For example, let us say you are Web surfing. Doing so moves a bunch of data in the background, and with every mouse click, data comes in and data goes out. Remember that the clone software first compares the main drive and ALL its files with the same on the clone. This comparison allows the software to learn and create its own "list" of what is to be copied from internal to external. The comparison process is done first and takes time, even before doing the actual data backup, sometimes as much as 30 minutes and maybe more, all depending upon how many files there are, all of which are scanned and internally catalogued by the backup software. Once the "list" is complete, only then does the software begin copying to the clone and erasing from it what is now gone from the main internal drive.

If you are working or playing during that critical cataloging time, the software makes what becomes an inaccurate list. Then, when attempting the copy activity, it can no longer find some of those files because now they are gone. The backup software reports these as errors during backup, which, unless you know what's happening, will cause you to think there are genuine issues with the backup, when in reality, nothing is wrong. I want there to be NO reported errors. Don't you? Still, even with these bogus errors, you have to check things by starting up from the clone and doing the verifications as specified. And what if the reported errors are genuine? You'd never know, even when looking at the error logs, because each reported error looks like so much gobbledygook, understood by only a well-experienced computer geek. Avoid all of this by following my advice, please! Just a few reported errors are likely not a problem, but check that clone, and check it with regularity! IF your computer will not start from the clone, there IS a problem.

In these cases, start anew by reformatting the clone and doing a fresh backup. If there is still a problem after a new, fresh cloning, contact tech support by looking up the issue in their online knowledgebase or by attempting to call tech support, but ONLY with Intego, as the others offer only email support. For the record, I have NEVER had the need for Intego tech support nor have there been issues with any family or friends using the product that were not fixed by following the above advice.

This technicality is also one reason that, for most users, Time Machine is going to be the way to go for *local* Mac backup. The downside is that a Time Machine drive is not bootable. The computer may not be started from a Time Machine drive, so in the event of an internal, main drive failure, getting back up and running is time consuming AND there is no immediate, temporary remedy that can allow work to go on until the drive is replaced. I hope you can see why I like this redundant solution.

Remember that whatever may be the means of achieving local backup, there must be a connection to that backup drive. Are your palms becoming sweaty at the thought of something so foreign, so seemingly geeky? I mention this not because of a desire to introduce an idea out of the comfort zone for so many readers, but because I want to offer solutions that work for almost everyone.

Please think about the possible *inconvenience* of doing a local backup. If inconvenient, then any solution is less likely to be adopted. And if there is no backup then so, too, there is no protection. I want each of you to take this seriously and to protect your data from loss. That's the only important point.

If you have a desktop computer, connecting an external drive via a cable is not going to prove difficult or inconvenient. But if you have a laptop, you know you are not going to keep another drive connected for continuous backup while moving about with the computer. You know you are not likely to carry around that added drive, nor should you. I want to limit your exposure to failure, so carrying around your backup is just not smart, especially if your laptop is stolen and the backup drive just happens to be in the same bag as was the stolen laptop. Now, maybe this is making more sense!

The great news is that updating a Time Machine backup is not a time-consuming task. If away from your base of operations, be sure to send *emailable* files to yourself as a temporary backup. Then, upon return, connect the Time Machine drive to the Mac laptop and click Backup Now in the Time Machine menu at the top of the Mac's screen. You'll see that it does not take very long to catch up and stay current.

Is there anything similar for Windows computers? In a word, no. Neither Time Machine nor externally bootable cloning is possible in the Windows world, but the above-recommended **Rebit 5** comes closest, even without being capable of external booting, so there is no way to spontaneously test effectiveness. You'll just have to trust it and have faith that all will be OK if and when it is needed.

I know your head is spinning from all this ponderous and boring information. Maybe, just maybe something is sinking in. Understanding the ins and outs of computer backup is confusing, and I hope to have helped readers to have more than a basic understanding of the topic. Yes, I have indicated a preference for what I believe is the ease of creating and using local backups created for Macs. It's just easier.

When it comes to Windows PC backups, there are so many choices, yet I have chosen to highlight and recommend the Rebit 5 product. If you are comfortable with another local plan, use it! In a further effort to understand the intricacies of Windows backup, I asked about Rebit's methods and reasons for doing as it does. You already know I like the full and complete backup plan. Here is why Rebit offers what I like: easily and transparent to the user.

A Rebit spokesperson was kind enough to respond:

Rebit is optimized around the idea of reverting the system hard disk to a past point in time. The point in time can be selected from a menu of past days such that a virus infection or other problem can be undone. Once the initial full backup has taken place, subsequent backups take a very short time and require only a small amount of data, which allows Rebit to retain a larger history of "Recovery Points" and file copies than traditional backup.

Then, I persisted, asking about the generally different ways Windows users have of performing local backups. Rebit co-founder, Dennis Batchelor, provided further insight:

Typically, the Windows community has the choice of two types of backup. The first is file-level backup where a utility basically goes through and copies the files from a pre-configured set of folders. Lots of backup products operate this way, and most of them fail to copy any file that is currently open. I tend to keep my Outlook email application open all the time, and I would be pretty upset if my Email file were not backed up. The other main drawback to this type of backup is that the operating system is not protected and cannot be restored in the event of a hard disk failure or virus attack.

The second common type of backup is *image backup*, where full backups consist of a complete, byte-for-byte copy of the information stored on the hard disk drive. As you might imagine, this can be a lot of data, so these are typically only captured once per week. During the period of time the image is captured, most of the computer resources are utilized, preventing the normal use of the computer during this time. Then, an incremental backup would be performed on days aside from the full image backup where only the areas of the hard disk that are changed get copied. The main drawback to this type of backup is related to accessing the data stored inside one of the incremental backups. In order to get a file from one of these backups, the software will dynamically merge the full backup and subsequent incremental backups to create what looks like a browsable folder structure. This procedure can take several minutes in some cases.

Rebit is an optimal blend of the two types of backup described above. It is optimized at the file backup level for quick and easy access to backup files. But, it also stores several days of Recovery Points (micro-images) containing data necessary to reconstruct a system hard drive. By minimizing the data necessary for a Recovery Point down to the file table and a few other pieces of data, the process takes only a few moments and allows normal usage of the computer. By running in the background and monitoring for periods of inactivity, the Rebit backup goes unnoticed yet keeps all your important files backed up throughout the day.

Local backups have been covered for Windows PCs *and* Macs, as well as recommendations for Mac antivirus prevention. What about Windows computers? My recommendation and what I use on Windows computers is the paid program called **ESET NOD32 Antivirus (eset.com)**. Their **Smart Security Suite** is, in my view, overkill and not needed if all my other advice is followed, including the use of a router as explained in the chapter called *"Computers – Laptop or Desktop, Windows or Mac, or Maybe a Tablet."* There are numerous Windows antivirus programs you may like or would like to try, and if you are so inclined, go for it. I use and recommend NOD32 because I know it does not slow down the computer AND, from experience, it does the job *for me* and for everyone I know who has taken my advice. No malware has gotten past this software on any computer in my sphere of users. What more can be asked of it? If you like and use another one, do as you wish.

What about free antivirus solutions for Windows PC users? Microsoft provides FREE antivirus protection called **Microsoft Security Essentials** (Microsoft.com/enus/security_essentials/ProductInformation.aspx) and is available for XP and later Windows versions through Windows 7. Other free solutions exist, as well, so search using FREE ANTIVIRUS for the list. OR, find them all, free and paid, on one convenient page with links at **microsoft.com/windows/antivirus-partners/windows-xp.aspx**.

Where will the local backup be stored? The best recommendation is in a fireproof safe! Though not practical, it is the smartest idea. College students should store their backups (on a portable hard drive, of course) in a locked drawer if in a dorm. If the backup is taken out and accompanies the laptop, the backup can be lost or stolen as easily as can the laptop!

This suggests that local backups are not sufficient for proper protection. Off-site backup, in addition to local, completes the picture. One plan provides that multiple hard drives be employed and rotated, with one kept, perhaps, at the office or at a nearby friend's or relative's and one at home. Alternating, switching them between both locations with each backup assures that if anything happens to one of them what is lost is only one generation of backup behind. It is easy to see that the idea of multiple local backups can become quite cumbersome, impractical, forgettable and even regrettable! No one I know follows this plan, but you have to admit, it does seem like a good idea.

Here's another thought to brighten your day! The backup, if in the wrong hands, just as if your computer was to fall into the wrong hands, can have its data accessed by unsavory characters, your personal and private data now out there and out of your control! Scary, isn't it! Even if your startup is password protected, bad guys might still get in. I have no simple solution for this dilemma, but it is worth noting. Sorry. Some will say you should set your computer, Mac or Windows, to encrypt all data, but this can affect performance. I know no one who does this, by the way. Keep the local backup safe from theft to the best of your ability.

The last piece of the puzzle is off-site backup we'll call *in the cloud* backup. This one is pretty simple in concept. For a modest annual fee of about $60, users can send away data for backup to a company's secure servers. Yes, secure. Data is encrypted when sent up to them and stored in this protected fashion. This is very good!

There are many companies vying for your backup business, but I will recommend one that I know and trust, though you know you are free to do your own research. **Carbonite** (**carbonite.com**) is a leading and recommended backup company providing service for Windows and Mac users alike, offering unlimited backup for individual computers for one fee, currently $60 annually. They also have plans for small businesses with multiple computers at additional cost and with limited amounts of data backup.

How does Carbonite work? There is a free two-week trial available, so why not give it a try? After downloading their software, users configure the "client" and Carbonite does its work in the background, not slowing down the user. The software knows when the user is taxing the network, downloading or uploading. Carbonite waits until things are less hectic and resumes its upload automatically, any time the computer is online.

Whenever the user creates a new file or changes one that already exists, Carbonite automatically encrypts and sends a copy of the file to Carbonite's data center, to the user's account. Files are secure – not even Carbonite can see them! They are protected with your password. Not only can the user retrieve and view files at any time, anywhere, to restore one or more files that become lost from the user's computer, but users can also see backed up files on any online computer, iPhone, Android phone or BlackBerry. In this way, Carbonite is much more than just a backup service.

In the event your hard drive fails, all your backed up files are safe with Carbonite and retrievable. Systems redundancy assures protection of your files; they have more than one encrypted backup of *your* backup.

Now, the downside of ANY cloud-based backup. Though users may backup unlimited amounts of data to individual accounts on Carbonite and some other services, there are practical limitations. Cloud-based backups rely upon the user's upload and download speeds. The upload speed is what is important for sending data TO the cloud server. Most modern computers have 500 GB or even 1 TB (terabyte) or larger hard drives. These are huge repositories of information, even though the drives are likely not nearly full. If a user has large amounts of music, photos, videos, personally created documents and more, say, to the tune of 300 GB or more, it is simply impractical to back it all up in the cloud, in addition to all the system data – the operating system and settings, and all the applications. NO online backup is designed to do a complete backup the same as are local backup schemes described above. You will not be allowed to back up system files or applications to Carbonite or most other cloud-based backup services.

Online backup companies rely on users saving personal info; music, photos, videos and documents. This is the data that cannot be recreated.

Apps, operating systems and the rest are somewhat generic and can be restored from scratch. In the event of a catastrophic loss, it will take hours and hours and hours, perhaps more than one day to download all that info back to a new computer or a new drive in an existing computer. For this reason, cloud-based storage is best used as a secondary backup, and as backup that is available wherever, whenever for certain, limited numbers of files. Despite this limitation, cloud storage is an important part of an overall backup strategy and should not be dismissed. The slower your Internet service, the longer all of this will take. It would not be uncommon for the first backup to take more than *one or two months* to complete.

To be clear, both local and online backup are a must!

I also know some users who create backup DVDs of photos and files as there becomes enough to fill one or more DVDs. I find this highly impractical, but you might feel differently. This method also does not back up the computer operating system, settings, updates, applications and other essentials. It can, however, be a way to provide a redundant set of backed up data that can be an additional safeguard in the event of calamity. Just don't rely on this kind of backup as primary.

Through it all, there is yet one more puzzle piece I rarely see mentioned, though it is critical and crucial for any user taking the first step in implementing a good and reliable backup strategy. You know that all hard drives fail, right? This includes ANY local backup you may perform. Yes, *THAT* hard drive. **Just as your main internal drive can and will eventually fail, so, too, will the drive used as a backup**. Don't worry about it. Just be smart! Depending upon the drive selected, it may come with a three- or five-year warranty. KNOW the warranty on the drive you buy for backup! The best advice is to replace that drive, the backup drive, in no more than five years, perhaps three years, depending upon your level of comfort with the whole backup process. I can only tell you what I do, and that is to replace backup drives every **three years**. I sleep better doing this.

The drive's maker will replace any drive that fails under warranty, but the *data* on the drive WILL NOT BE RECOVERED FROM THE FAILED DRIVE. The warranty covers only replacement of the drive itself. Do you see where I am going with this? YOU must manage all of this! Mark your calendar for a reminder three, four or five years after purchase to replace the local backup drive with a new one. The good news here is that a 1 TB drive purchased new today will cost considerably less in three, four or five years. It is likely that in the years ahead, a 2 TB drive will cost as little as half what a 1

TB drive costs today. Even if the drive seems perfectly fine, don't trust it indefinitely, and certainly you do not want to discover your backup is no longer working at the same time you experience a failure of the main drive. I am not trying to be an alarmist. I AM trying to be a realist, however.

Yes, this means throwing away, hopefully, a fully functioning drive. Get over it! Decommission the to-be-thrown-out drive by doing a low level format, also known as a secure reformatting of the drive. Why bother? If there is any personal info on the drive and it is simply thrown out, someone, anyone can retrieve the drive and recover, perhaps, your passwords, Social Security number, addresses and credit card numbers and passwords to your accounts, plus all the email addresses of your contacts. Can you say **Identity Theft**?

Basic reformatting or a simple erase procedure does not remove the data. The process does no more than tell the hard drive that all areas of the drive may be used as if new, overwriting the original data. DON'T DO THIS, as it is not a secure method and the data is recoverable!! Secure erasure means NO ONE is likely to be able to get data from the drive, or so I am told. Surely, however, other than government forensic scientists, doing what follows is as safe as it gets. Do this ONLY after you are certain, without doubt, that the NEW replacement drive is working fine to the best of your ability and that the NEW backup is tested and complete. Wait a week or two, maybe longer, to erase the old external drive. Once the drive is erased as specified, all your data on that drive *will be toast*!

An easy way to *securely* erase an **INTERNAL Windows** drive is with FREE software called **Darik's Boot and Nuke (dban.org)**. Download and follow instructions. Use this when the computer is to be decommissioned OR when the INTERNAL, main drive is to be replaced. The software is NOT for external drives. Erasing EXTERNAL drives, such as your backup drive in need of replacement is as easy as following the steps from a FREE program called **Eraser**, with instructions and a download link at **secure.nd.edu/disposal/eraser.shtml**.

Again, do not erase an external backup drive until a new one is in place and has been shown to be healthy. Hard drives should be disposed of as e-waste, so learn the procedure from your local government.

Mac users have it almost as easy. Start up the computer whose internal drive is to be toasted using a Mac OS install disc OR with one of the original discs that came with your Mac. This is done by inserting the startup CD or

DVD, letting it load and then restarting with the "C" key held down. This tells the Mac to start from the disc in the CD/DVD drive and not from the internal hard drive. You'll know the Mac is started from the CD or DVD by the look of the desktop when it is finished starting up. It should look as if you are going to install the operating system again. You might need to click though a screen or two until you see a menu at the top and to the left of the screen, indicating **Utilities**. Only when you see this, click the drop down menu to the choice of **Disk Utility**. Select it and let that application load.

Select the level above the **Macintosh HD** at the left, and then click the **Erase** tab at the right. Once there, select **Security Options** at the bottom and then choose a medium to high level of security on a new Mac. On a pre-Lion OS X Mac, choose **Security Options** and then select the **7-Pass Erase** option, which will take up to a few hours on a 250 GB drive or close to a full day on a 1 TB drive using the **35-Pass Erase** option. Click **Erase** and **OK**. This is heavy-duty protection! When finished, quit the utility. If it is the goal to erase this hard drive so the Mac can be repurposed for another user or disposed of as e-waste, after this process it will be ready for what's next. If this is being done so that the drive can be removed and disposed of, replaced by a different drive in the same computer, the erased drive will be ready for e-waste disposal.

When finished, quit the **Disk Utility** and restart the computer holding down the **eject** key. The computer should spit out the CD or DVD, and it will not start up if it is the internal drive that has been erased. In this case, once the optical media is ejected and retrieved, hold down the power button until shutdown.

If there is an external Mac drive to be erased and disposed of, the procedure is similar, except that the Mac may be booted normally if it is still alive, so to speak. If not, start using the CD or DVD as above, holding down the "C" key. Once it is started, connect the external drive, then open the **Disk Utility** as above, select the upper level of the external drive and perform the erase function as above. When completed, that drive is ready for e-waste disposal and can simply be ejected, as would any attached drive. Don't bother trying to format it if a warning appears after erasing the drive.

Finally, and I know you are relieved that this ordeal is just about over, what about theft protection software? This is a popular option for students away at college. It is an area of interest in which I have done little hands-on testing. I would, however, be remiss for not mentioning it as a part of the total protection package worth exploring. I'll leave you on your own here, but there are some top-rated referrals I can offer! Products in this category offer services that include remotely, stealthily "phoning home" when the computer is reported stolen. Numerous snapshots can be taken with a laptop's built-in Webcam and automatically sent to the registered owner's email without the knowledge of the thief. In addition, the software may be able to track and send the laptop's location so the owner can alert authorities to its whereabouts for speedy recovery. Another function may be to remotely, securely destroy data on the stolen laptop. There are other capabilities, so if you are interested, look into the following software:

Prey, at **preyproject.com** is **FREE** and may be used on both Windows PCs and Macs. The company also offers enhanced paid services. **IPsneak** (**ipsneak.net**) is a **FREE** product for Windows only. **Locate My Laptop** (**locatemylaptop.com**) is another **FREE** Windows-only product that also offers upgrades paid services. **Absolute Software** (**bit.ly/so5zB1 – shortened for your convenience**) offers a well reviewed paid service called **LoJack for Laptops**, for Windows and Mac laptops.

CHAPTER 6

HANDS-FREE PHONES

SOLUTIONS FOR MOBILE AND HOME CORDLESS PHONES

I have no quarrel with hands-free laws for mobile phone users. I DO have issues with most Bluetooth hands-free solutions, however.

I've never used a Bluetooth headset or heard others using one while on a call with me that sounded as good as without the device. Some are worse than others. There are many ways to describe the poor audio quality. It is as if these things get a pass on audio quality in exchange for their convenience factor.

For me, one of the tests has always been how close to natural they sound. The truth is, none have a natural sound. Some are purported to be good at blocking ambient noise, *noise cancellation*, while others are style oriented. My experience and that of most reviews comment upon poor sound quality.

Voice quality is always a compromise, admittedly, something not as good as the sound from holding the phone to one's ear and just speaking, but not all phones have great sound and none do a good job natively rejecting ambient sound (noise cancellation) to the other side of the call.

All the top rated Bluetooth headset models are quite small, and all have attractive extra high-tech capabilities. This could include accessing voice command features built into the headset and in the connected smartphone, special touch combinations that trigger actions or other non sound-related extras.

These things are marketed for their *gadgetry* tech features more than for their functionality as good, quality sound devices. Still, I know they have their purpose, including as a status symbol for the wearer. At about $100 - $150 for most top-rated devices, it is not uncommon for some consumers to get whatever is the latest craze simply to keep appearances and to demonstrate one's *coolness*. When another well-reviewed model appears, many consumers follow the herd and buy that one, and the pattern repeats when the next must-have headset is a top pick with reviewers.

Coolness factor aside, there remains three principal issues ever-present in Bluetooth headset reviews – comfortability, sound quality including noise cancellation, and battery life – shared among all popular and well-reviewed such products. I've never experienced a device with exemplary performance in all three areas. Something, often quite a bit, misses the mark.

If the underlying mission is for a BT headset to comply with hands-free laws, then it's mission accomplished. If the purpose is primarily as a *convenient* way to converse without having to hold phone-to-ear, success!

HOWEVER, as a tool by which to conduct hands-free calls with the best sound quality, noise cancellation, all day long wearing comfort, and without batteries that limit useful operating time each day, they all fail miserably. And for some users, their increasing complexity is vexing.

I recognize the convenience factor of having an unencumbering wireless device in one's ear with which to carry on phone conversations, and in circumstances in which this wireless link between the phone and ear is the most desirable characteristic, Bluetooth headsets meet the objective well. I get that.

I want to suggest an alternative to the BT headset, what I use and find excellent in most daily circumstances. This alternative offers exemplary sound quality and noise cancellation to both sides of the conversation. The other party will probably not suspect the user isn't using the phone's built-in microphone and will not hear the bulk of any extraneous noise, even loud noises. The only drawback is that it is **a wired device** and not very sexy. It is also a single-purpose device for making calls with one's hands free, without any additional built-in features.

My choice and recommendation is not a new product, either, and that is part of a drawback perceived by some – its lack of sexiness, that it is not hip and new, hot and trendy. It just gets the job done, better than anything else of which I am aware. MY objective in hands-free calling is to have the best audio experience, with all day long wearing comfort a big positive. And I like that I never have to recharge the headset or deal with software!

If you can get past the pedestrian nature of this product, satisfied with its sound quality and ability to reject background noise and wind, you're going to love this wired headset. It is a winner for voice commanded mobile phone activities on iPhone, Android phones, Windows Phone devices and any others not included here. Oh, and it is also the perfect companion to use along with your headset jack-equipped cordless phone and headset-compatible office phone. Use it, as well, with Skype and any dictation and other voice-related and voice activated input to a computer using available plug adapters. It can do SO much and do it well!

I use mine while in the car any time I am planning to make or receive calls. It lives in my car and is easy to connect. On a short or long drive, it is the most comfortable mobile phone headset imaginable. Drive without sending noise to the other party, even in a top-down convertible or with the air conditioning blowing at one's face. While in my office, I can wear this comfortable, preferred headset all day, every day without discomfort or batteries!

Bluetooth headsets don't sound better while driving than when walking on the street. Visor-mounted vehicle Bluetooth speakerphones are another option with their own attendant drawbacks. These include poor sound quality and poor noise cancellation properties. In addition, there is a total lack of privacy with such devices.

Many vehicles are equipped with Bluetooth technology for built-in hands free convenience and compliance with hands-free laws. Unfortunately, many of these systems are not easy to use and present less than ideal sound quality. They all come at a significant additional cost. My informal and anecdotal survey of friends whose cars have this built-in capability indicates the systems are forgettable and regrettable soon after the new car honeymoon is over. These friends would have opted out of this expensive extra had they known of the system's shortcomings before purchase and if they had the opportunity to buy the car without it. Even on high-end systems, none of which I am aware are good performers when speaking at normal, intimate conversational volume. It seems they all need the user to speak at elevated sound levels for optimal performance. Most do not feature a level of acceptable noise cancellation and if voice activated leave much to be desired. Some vehicles include voice recognition along with Bluetooth. I've tried several systems. When at a stop, most will function at least close to adequately, so long as the vehicle is stopped, the radio volume is off or auto mutes upon voice control activation, and the windows are up. Once in motion with and without the windows open, their ability to recognize and execute a voice command is quite poor. This is so unnecessary! Moreover, what a waste of money!

It's time for the big reveal . . . The winner is . . . **theBoom** (**theboom.com**). Never heard of it, have you? They have been around, popular in certain circles for more than 15 years. I've used and recommended them for more than 10 years. What I particularly love about **theBoom** headsets, in addition to their sound capabilities, sound quality and build quality, is that they never become obsolete, never require a software update and do not use batteries of any kind to accomplish their magic. Owning one or more is an investment providing many years of benefits!

The products live up to the company's registered motto – **Whisper and be heard™**. Period. Imagine being able to speak in *soft normal tones*, even when out and about in the noisiest situations. It seems unnatural to speak softly when in the midst of loud background noise, but **theBoom** products are THIS good. We've all been in the presence of mobile phone users who speak as if they are the only ones in the area and that they must speak very loudly in order to be heard on the call. Isn't that annoying! Using **theBoom** headsets, users may speak, in fact, are encouraged to speak as if in a quiet, exclusive restaurant conversing on matters of utmost privacy and import. Do you have that concept clearly in your head? Now, right now, say something in this almost hushed tone. I know it is so contrary to everyone's experience. THAT is the voice users may employ and enjoy with complete success using **theBoom** under all circumstances. Try *that* with ANY other headset. I dare you, and especially if the noise drowns out your own voice in your ears. The other party WILL hear you and you will be understood with **theBoom**!

Visit **theboom.com** and click on the **Consumer** link. Then click to play the demo video. You'll be hooked after watching this actual demo.

With unique and patented technology, all **theBoom** headset microphones cancel extraneous noise by their physical design; hence, no power is needed to do what they do so well.

If your phone has speech recognition or other voice control capabilities, theBoom headsets are fully compatible with all these features even while driving at high speeds, even with the windows or convertible top down, even with the air conditioning blowing at your face or with the radio on, a feat no built-in voice recognition system on any car at any cost can accomplish. Background noise disturbance will not be transmitted to the other side of the conversation. All the while, you may speak in a softer, normal voice, as if there was no background sound that would otherwise drown out your voice.

Using theBoom presents a more professional YOU to all callers by virtue of the excellent sound of YOUR voice without background disturbances.

Three models make up the universe of possible theBoom products ideal for *most* consumers.

theBoom O, base priced at $110, is perfect for over the head all day comfort, ideal for use in the office, at home or whenever this style may be most comfortable for repeated all day wear.

theBoom V4, base priced at $150, is what most users would appreciate for use in the car and walking about in or out of the home or office. Its contemporary, over the ear design and appearance looks "normal" and won't draw undue attention.

theBoom E, base priced at $299, is a customizable model for the enthusiast in offices, at home or in the car. Its behind the head and over the ear headband and curly eartube with in-ear sound offers the best audio quality without the having huge "cans" found on over-the-ear style headphones. On this model, sound emanates from a tiny, high quality speaker built into the headband, and is sent to the ear through the hollow, curly eartube. Users may choose from left or right ear models, or both ears. The eartube is reminiscent of a similar look you've seen on TV when there is a curly eartube behind the ear of on-air news or other talent. Law enforcement and military personnel use the same technology for personal communications.

How much have *you* spent on phone and computer headsets? Wouldn't you like to pay once for many years of performance and enjoyment, not needing or wanting to buy something more high tech, somehow better in the near term? This is both a blessing and a curse for a product like **theBoom**. They are already nearly perfect, so where is there to go? Newer? Better?

They are already there. Regardless of which of the three may be selected, they all have the same microphone technology in their design that allows them to perform so perfectly in all situations.

Just so you'll know you are not alone in learning that **theBoom** is the best at what it does, note that **theBoom** may be found and is loved on the floor of noisy stock exchanges, in call centers, in auto racing, in aviation and in other high-noise environments.

Not only do I enjoy finding best-of-the-best products and technology most do not know about and sharing this information with Mr. Gadget® followers, I also like working with nice people. It is another plus in my dealings with this California San Francisco Bay area company. These are among the nicest people in the industry and have outstanding customer service!

CHAPTER 7

HOME PHONE SERVICE

INEXPENSIVE, EXCELLENT QUALITY, CORDLESS PHONES, BATTERIES

I am NOT entering into the debate over whether it is worth the cost to maintain a traditional home phone or to jettison it in favor of only a mobile phone. That is up to you!

However, to everyone who, as I, and for whatever the reason, wants a "home" phone, have I got money-saving ideas for you!

What do you pay for your current home phone service? I'll bet it is at least $20 to $30 per month, if not more, with additional services raising that monthly charge even further. **How about as little as $30 _per year_?**

Here I will report on the five best and tested alternatives I've found, with certain reservations as noted for each.

- magicJack
- magicJack PLUS
- netTALK DUO
- Vonage

- **Phone Power**
- **Cordless Phones and Rechargeable Batteries**

The way to beat the high cost of phone service from the traditional phone companies is to use the Internet! You do have Internet service (high-speed DSL, fiber or cable, not satellite), don't you? Calls routed over the Internet necessitate that the user have reliable, mostly up-and-running Internet service. In my experience, it is also important, if the speed of your service is not very fast, that other computer activity be minimal while on a call. If not, call quality may be choppy when using some services as compared with others. More on this later.

From its public introduction in late 2007/early 2008, **magicJack** (**magicjack.com**) has been the low price leader at [now] about $40. It is simple and basic, yet offers several added features, such as E911, voicemail to email, call waiting and call forwarding, and more. **See all features at the FAQ page** on their Website.

There is something new! Whether you are already a magicJack user or soon will be, once signed into the user's my.magicjack.com portal, there is a link to number porting, a long-promised feature. If already a magicJack user, visit the link to see if your area is covered in order to take your own phone number to your magicJack. Not all area codes or prefixes are covered, but it's worth a look. At this writing, the cost to port a number is $20, but you will find the final word on cost, if different, as you investigate it on their site.

I've used magicJack since its introduction and have also provided the product to others to gauge their experience, both in the US and in foreign countries. This is the consensus – higher speed Internet service helps the experience to be as seamless as traditional home phone service, but basic high-speed Internet service is all that is generally needed, heeding the quality caveat above. By basic, I mean service with at least 128kbps *upstream*. After all, when talking, the other end hears the result of your sound being sent UP to them. You hear the other end by using *your* DOWNLOAD speed (*to* you), which is usually considerably faster, at least about 768kbps. Test your speed by visiting **speedtest.net** and be sure there is no other known online activity while performing the free test.

Available through the company Website and online retailers as well as inside many retail stores including BestBuy, Radio Shack, Sears, Office Depot and Walgreens, the original $40 magicJack relies upon direct connection to an always-on computer. That's right, the "host" computer must be ON and awake, not sleeping or otherwise not at full alertness. It uses the computer's USB jack and the computer's Internet connection to complete the circuit, so to speak. Whole-house service is accomplished by using single or multiple handset cordless phones, the base of which is connected to the magicJack device's standard RJ11 phone jack. A single hard-wired phone also works just fine. It *is* that simple, in my experience and in most cases.

However, I also know of some users who were not able to get it to work and complained loudly. If your computer is of modern vintage, meets the minimum system requirements and otherwise performs without you cursing at its shortcomings, there is a good chance your Windows PC or Intel Mac will readily accept magicJack just fine. Again, heed the Internet speed caveat above and PLEASE don't even try to make it work with dial-up or Internet via satellite service. That said, here is how things generally go . . .

Plug magicJack into a USB port, automatically launching the on-board software for installation on your Windows XP/Vista/Windows 7 PC or Intel Mac. Once installed, users continue set-up by selecting a convenient area code and prefix. Not all US areas are covered with local area codes and prefixes, but consumers are encouraged to learn what is available through the FAQ/Knowledge Base link at the company Website (which can be checked BEFORE purchase on their Website). If your state and local area are not currently available, choose anything you like. For a fee of about $10, you can switch at a later time to a new number if a local area code and prefix become available.

Then, after entering the user's address (for E911 service) and completing other simple setting screens, pick up the phone and call *any* US or Canada number without added cost for one year.

Annual renewals are $29.95 per year or less with a multi-year deal as low as $100 for five years!

What's not so hot? Nothing that I have encountered. However, as noted, there have been many complaints. I've read page after page of them. Most areas of dissatisfaction come from unrealistic expectations, in my view. Things such as customers not knowing that the device needs to be connected to an always-on computer, or that there is no live tech support available, that it is only via online chat. Everything has worked smoothly here and with testers I know. Service renewal has also gone smoothly. All my testers have at least basic DSL or basic Internet service from their cable company. Some, like me, have very fast fiber optic Internet service.

It is only when those with basic high-speed service are heavily taxing their capabilities that magicJack service may become choppy, that voices cut out and it becomes difficult to hear for either or both parties. In that case, knock off downloading or other activity that may also be in the background.

The company has taken heat for its cheesy marketing and Website. It *is* cheesy! My experience has been positive. What have you to lose? magicJack offers a 30-day return policy (not including shipping costs, if any). Note that this is 30 days from purchase or online order. DON'T let it sit for a year before trying it and then try to return it. That's 30 days and only 30 days. Follow the rules and all should work out for most users. If purchased locally or from an online source other than directly through the magicJack.com Website, ask the retailer about their return policy.

Services from magicJack in addition to the free calling include these free extras – directory assistance, call waiting, voicemail, call forwarding, and caller ID (with compatible phones, of course).

Traveling?   Take magicJack along, perhaps with a plug-in corded phone, even internationally, and call home to the US without charge, so long as your computer has available high-speed Internet service. If not a corded phone, use the built-in magicJack softphone, that is, an on-screen display of a dialing pad and your address book (stored online), and use an external microphone or the laptop's built-in microphone, if equipped, along with the computer's built-in speakers or plug-in headphones.

Plain and simple corded phones are readily available for under $15 online and from retailers including Radio Shack and Best Buy. Some are listed for well under $10.

Note that E911 – the 911 emergency service tied to Internet-based phones and mobile phones – will not work internationally, so DON'T try it when out of the US. If magicJack is to be used away from the current registered address for an extended period, change the registered E911 address in the settings. **This same mobile E911 rule applies to *any* portable Internet-based VoIP service.**

Speaking of international calling, magicJack rates to most destinations outside the US are among the lowest I've found, but they vary widely, so please investigate before purchase if this is a possible selling point for any of the listed services/products. It's easy. Rates are available for review at the FAQ/Knowledgebase link on the magicJack Website and easily found on the other suggested alternatives' Websites. If there are regularly called numbers in other countries, check rates and compare before signing up for service.

There is something new at magicJack, well, actually two things! First, the company is rolling out the long-promised capability of taking an existing phone number along and NOT having to get a new number when buying magicJack. This is called "number porting." Users may begin service with a new number and then investigate whether it is possible to take along the old number through the my.magicjack.com portal. Click on My Subscriptions, then Details under the link to "Transfer Your Phone Number to magicJack." $20 is a real bargain if you want to keep your existing phone number. Potential customers may check availability of this service for an existing number by clicking on the FAQ link at the bottom of the company's main page as listed above. Once at the FAQ page, enter the word 'port' without quotation marks. One of the links found will be "Can I transfer or "port" my current land line telephone number to magicJack?" Enter your number and email address and within a day or two, you will be emailed an answer.

Why all of this? magicJack can only port a number that is in their recognized, authorized service area, of which there are many. Most, but not all areas of the US are covered. The US is a big country. The backup plan is to check out not only number porting but also available area codes and prefixes, also at the FAQ page, searching on the words "area codes" (without quotes).

The second new *thing* from magicJack is the $69.95 magicJack PLUS. it is everything the original magicJack is, only more flexible. It *can* be connected to a computer, as *must* be the original, OR it *can* be connected to a router, bypassing the computer, except for initial setup. The most criticized feature – that magicJack must be used with an always-on computer, is no longer the case with PLUS! This configuration may also provide a tremendous energy savings over having to maintain an always-on computer in order to use magicJack.

E911 is a bit different on magicJack PLUS, and better. On the softphone (on the computer screen), there is a 911 dropdown, allowing multiple US addresses to be entered, perfect for the mobile user who takes it along on the road. Another obvious use would be for someone with multiple residences. Toggle between locations as needed!

The PLUS device also includes number porting possibility as explained above. It usually takes from a few days to a week for your existing phone company to release your number to magicJack for them to implement it as your new magicJack PLUS number.

Oh, and software is automatically updated on both magicJack devices as needed.

How well does magicJack work? Everything's just fine in my tests and those of my "testers," including two overseas participants. In those cases, I set up magicJack with a US number and sent it to participants in two foreign countries. Each of them now has a number in a US city and can make calls to friends and family here in the US as well as receive US-based calls as if local, even though the call is received thousands of miles away in foreign countries. This is a thoughtful way to provide a free to inexpensive call method for anyone you know in another country. Businesses, take note, as well. Other alternatives below may be even better for businesses!

Both magicJack and magicJack PLUS may be used on multiple computers. Once registered, simply plug in the device to any compatible computer, assuming Internet access is available, and it should work after the simple software setup has completed on that computer. No re-registration is necessary. In the case of the magicJack PLUS, connecting it from one router to another, once it is registered, additional registrations should be unnecessary. Plug it in to other routers, wait about a minute, maybe two, for it to communicate with *HQ*, so to speak, and say, "Hi, I'm here! Say hello back to me," plug in the phone and call out or receive incoming calls! Switch between Windows PC and Macs with ease!

Our experience with magicJack Plus is equally positive. Connect it, set it up and it works, with sound quality no one has complained about – except as explained above in the case of slower Internet service with too much going on. Regardless, it is worth the effort to give it a try. You cannot beat the price and performance for the cost of either regular magicJack or magicJack PLUS.

Search online for availability of magicJack PLUS in addition to the company Website.

Newer still is the new and **FREE magicJack APP**, available through the Apple App store for iPhone, iPod touch and iPad. Read the details at the App store!

Another device to check out is called **netTALK DUO** (nettalk.com) and advertised with features similar to the magicJack Plus in operation, at a cost of $70 and pre-dating magicJack Plus' availability. Service renewal is available for $29.95 per year. Our tests indicate good service with no major issues, even over minimal high-speed Internet speeds (except as noted above with magicJack), but **ONLY when this device is connected to a router.** Subjectively, audio quality was better using magicJack! netTALK DUO, too, was set up here and is now being long-term evaluated in a foreign country where Internet service speed is, at best, not very good. It is performing equally well as it did here in the US and was set up with a local (to me) SoCal area code and telephone number so we can call there as if the party is virtually next door, even though the reality is that the family is more than 10,000 miles away! It is connected through a router, always on and available, with a cordless phone connected to the netTALK DUO.

A deal-breaker of a problem occurred when we attempted to connect netTALK DUO to two computers' USB connection. Oh, the pain! Where magicJack just works, is easy to set up and then does its job without fanfare, as one would expect, netTALK DUO is the opposite. We, that is, I AND an accomplished IT professional tester, gave up after failure to get netTALK DUO set up and directly connected to our Mac and Windows PC. It was SO easy when directly router-connected and virtually impossible to get it to run, even to get it set up, using direct USB connection to computer.

After a frustrating 20-30 minutes each on a Mac and Windows PC, we simply gave up. We did not bother calling their tech support, which is available by phone, where magicJack's tech support is via email or online chat. Why? If we could not get it to work simply, easily and without needing tech support, I cannot give it an endorsement *when connected that way.* Isn't it important to NOT need help to set something up? One should not have to get help to set up a phone! And since two experienced and knowledgeable folks could not make the thing work, what chance has the average consumer? Note that our computers are the same ones, without any fancy or unusual configurations, used for all these other tests. The other products all worked, but not netTALK DUO when connected directly to the computer. Remember, it works fine when connected to a router.

What is the lesson here? netTALK DUO may be suitable for users who will NEVER need to direct-connect to a computer. Travelers take note! On the other hand, magicJack devices set up easily and quickly AND work well when connected via USB to Intel-based Macs and Windows PCs. magicJack is also less expensive to operate with its multi-year plan – five years service for $100.

netTALK DUO is available through the company Website and from online US retailers including **walmart.com, amazon.com, dell.com, pcrichard.com, newegg.com** and **buy.com**.

Voice mail retrieval on all three devices is available via pass code on the attached phones and from an entered code when calling in from the outside, typical of regular phone company voice mail services. Also, voice mail is sent as a .wav attachment to the email address of choice.

You may wonder how these three devices perform in the event of a power outage. After all, your old phone company hard-wired phone may work during a power outage. It's simple. All that is needed for uninterrupted service is a battery back up device, also known as a UPS (Uninterruptable Power Supply). With the desktop computer, in the case of one of these three devices, USB-connected to a computer, be sure the UPS has the computer plugged in, as well as the router, Internet modem and cordless phone, if appropriate. In other words, all the products that would fail without standard from-the-wall-power need to be plugged in to the UPS. Power out? UPS takes over, seamlessly."

You'll also need to calculate or have help figuring out the capacity of backup battery or UPS needed for your setup as well as how long you would want to be prepared with auxiliary power if regular power fails. Brands I recommend are **Tripp-lite** (**tripplite.com**) and **APC** (**apc.com**). Shop online through your search engine of choice for best pricing on the unit you want. Expect to spend less than $100 for a typical unit, perhaps under $50. I'd recommend, however, figuring need based upon providing at least one hour of power to the attached components. I like to have much more battery backup capability, however, as we live in earthquake country! If experience dictates that power is regularly out for longer periods, buy up accordingly. The same advice holds true for other Internet-based telephony solutions to be mentioned next.

On these or any similar services, if the connection to the Internet is interrupted (and there is no battery backup, for example), there is a failsafe plan. **magicJack**, **magicJack PLUS** and **netTALK DUO** all send calls to voice mail if service is down for any reason. If the call cannot go through, it goes to voice mail. Simple as that! Alternatively, if you also want to, on-demand, forward calls to another number, perhaps a cell phone, this is accomplished through the online device settings under call-forwarding. Don't forget to switch it back when needed!

If you want to keep your existing phone number, or even if not, you may also choose a more upscale provider. These provide the many other services of a traditional landline, yet they are still Internet-based, plus offering even more services all for a basic fee. Among many long-standing providers, most notably Vonage, there are others, too, to recommend. As this chapter focuses on best priced AND best performing Internet-based telephony solutions, Vonage doesn't make the cut only because it is more expensive than the leader. The same is true of most cable TV service providers, which also offer VoIP services to their subscribers.

Number porting is generally at no cost by these more upscale VoIP companies. A charge *may* be incurred from your current phone company, the one that has your current number, to release your number so you can set it up with your VoIP provider. It's not supposed to happen, though, from what I have been told.

I have used **Vonage (vonage.com)** and have no complaints about their service. My *only* issue is that their most economical unlimited service is $25 per month. Lesser plans, akin to limited minutes cell phone programs, are as little as $12 monthly for 300 outgoing minutes to the US , Canada and Puerto Rico. This may be a good choice for users who make few and short duration *outbound* calls per month. Note that incoming calls are always free. Visit the Website to see all their plans and the very long list of included services, much more than extra-cost services from traditional phone companies. NOTE that the Vonage Website may show a lower cost, but read the finer print, stating that this lower price is for three months only, after which the price goes up considerably.

One area in which Vonage shines brightest is for customers who make regular overseas calls in copious amounts. Some Vonage plans include unlimited calling to certain foreign lands, but the free calls are only to the country's landlines.

No matter which VoIP carrier is chosen, there may be a fee to call a foreign, overseas *cell* phone. A call to the UK, for example, to a landline may be included and FREE in some Vonage plans, but calling a UK *cell* phone over Vonage is currently priced as high as 34¢ per minute! The best service, recommended below, has most UK cell phone calls costing 6.3¢ per minute, with but a very few prefixes costing as high as 27¢ per minute for the call.

Beyond the plans that include international calling, international calling rates are also low with Vonage, though not as low as the next candidate to be highlighted. Check rates at the company of choice's Website!

Calls to Vonage subscribers whose adapters are not making contact with Vonage, perhaps due to a power outage, Internet being down or another reason, go to voicemail if user settings dictate, OR to a failsafe number, perhaps a cell phone that has been entered through online settings. Simple, tidy and neat.

Vonage voice mail *includes* attempted transcription of the voice message to text, sent in email, as a text message or both. Vonage allows multiple email addresses to be sent the voice mail notifications; a nice touch! The transcription attempts are often humorous. Again, Vonage service is generally excellent! Note that tech support is either via online chat or a call to someone that is likely in India or the Philippines.

For more than two years I have been using a worthy and lower cost alternative called **Phone Power (phonepower.com)**. They, too, have monthly plans without contracts, as well as monthly plans with commitments. Their best and recommended deal is this - $200 for a year with a second year free, which amounts to under $10 per month including taxes and fees for two years. The company says it is $8.33 per month, but this does not include mandatory taxes and fees, which I have included in the under $10 monthly figure. This gets you virtually unlimited calling to the US and Canada (up to a whopping 5000 minutes per month). Voice mail can be sent to email **and** a notification in a text message. **Phone Power's international calling rates are THE lowest overall of any company I've found.**

Service has been outstanding! See their Website for the long list of included services and other details, including 60 minutes of International calling per month (landlines only, of course). One incomparable extra provides a free second line for Phone Power subscribers (see Site for details). We use it with a second cordless phone set so there is always a phone available and ready for a second caller to call out OR to receive an inbound call by someone else at home. Cordless phones are quite inexpensive (see my recommendations coming up)!

All tech support for this company, unlike Vonage, is right here in the good old US of A. In fact, it is nearby to me here in Southern California. I've even visited their offices and call center.

What about faxing? For those *still* faxing and who have VoIP calling, NO VoIP provider easily accommodates faxing. The companies do not officially support it by guaranteeing that faxing in or out will work on your voice VoIP line. However, they all try to make outgoing fax work with instructions to set faxing to the lowest speed to start. Then increase speed until reliability suffers and back it down a notch. With these inconsistencies, faxing is something to be done only occasionally using VoIP. Incoming fax involves sharing the line with a fax machine. In the case of Vonage, for example, it can be hit and miss, but most report that it works, though it is helpful to know that a fax is going to come in so the fax machine can be set to manually answer the call before it goes to Vonage voice mail. Vonage also sells dedicated fax capability for residential customers using the Line 2 jack on their supplied adapter for a $10 monthly fee. Still, it is likely not going to be perfect.

magicJack and netTALK DUO's fax capability is this – If it works, great. If not, they do not officially support it. In my experience, outgoing faxing works most of the time, so long as the fax machine can be set to lowest speed and/or there is a VoIP setting that can be used to try and get it right. Still, there are no guarantees of reliability. Incoming faxes are more problematic, as they are with standard phone lines sharing voice and fax (in most cases). If you know the call is an incoming fax and if that line is shared with a fax machine, you must manually select to answer the call by the fax machine. See your fax machine instructions for how to do this.

Phone Power's approach to faxing is unique and better. When a fax comes in to your Phone Power number, don't answer, if possible. Let the call go to voice mail. The fax will be received in the TIFF file format (as an image) and sent to the email address of record as an attachment that may be opened as would any tiff-formatted document/image. If you *do* answer, hang up. The fax will likely be re-attempted in a few minutes (and this time do not answer). NO fax machine is needed for incoming fax messages with a Phone Power account! Outbound faxing is still hit or miss, with success most often the case in our tests.

If a call cannot be completed to our Phone Power device, we've set up a failsafe forwarding number through the Phone Power portal. If Phone Power is not working here, the call is sent to the assigned number, which is a mobile number. The caller has no idea of the problem and has no reason to know. The call goes through seamlessly!

Voice mail notifications go to one assigned email address and as a text message to one assigned number. As noted, Vonage allows multiple email addresses to be set. My workaround for this was to set up in my Gmail a forwarding address for any email coming from the specific email address from which Phone Power sends voice mail. This works just fine, so both my email address and *Mrs. Gadget's* get the notification!

I've covered all the basics, detailing what I believe to be the best candidates at the lowest cost. Not recommended are services from cable TV providers. This is for one reason – they are not the best value, not by a long shot. I am also wary of newcomers that may not be around for long. Sure, there is no guarantee that any service recommended here will be around as long as the good ol' phone company, but I will take my chances in view of the cost savings. Exactly how much less expensive could it get than the Mr. Gadget®-tested choices above?

Cordless phones are the phones of choice for most home phone users. After using most of the popular brands and some lesser-known names, I have arrived at a conclusion and recommendation – Buy cordless models that use ONLY standard, readily available AA or AAA rechargeable batteries!

From my own anecdotal survey among friends and family members, it is not usually the phone that fails, it is the individual batteries or battery pack inside that causes dissatisfaction. Batteries inside cordless phones are one of those out-of-sight, out-of-mind things. Users typically limp along with failing batteries offering minimal talk time until frustration reigns and the system is replaced. Have a look inside your cordless handsets?

If you find a battery "pack" consisting of plastic over-wrapped batteries, either with a metal contact tab or with a small plug to connect under the battery door, you are less likely to deal with battery replacement. It is just too much of a pain! And they are too expensive. Priced from about $10 to $30 *per pack*, it may be less expensive to replace the phone system than to replace one or more battery packs inside this type or cordless phone. Don't do this!

What if the otherwise fine-performing cordless handset(s) used rechargeable AA or AAA batteries? These are cheap and plentiful. **I use and recommend Panasonic cordless phones for their great prices, overall quality, reliability and durability, plus voice quality and features AND that they use these replaceable, readily available and inexpensive rechargeable AA or AAA batteries!**

In anticipation of problem avoidance, I chose (and recommend) a Panasonic DECT 6.0 multi-handset cordless phone system for use at *Gadget Central* just as I am recommending to you. It came with minimally powerful AAA rechargeables, Panasonic-branded, of course. I decided to act preemptively, immediately replacing those lower capacity 600mAh NiMH-type rechargeables with higher powered, longer running and inexpensive Lenmar rechargeable batteries. They are 1000mAh rated NiMH batteries for more talk and standby time between charges. I NEVER want any user to experience running out of battery power while needing to use a phone here. Ratings of 1000mAh are the max for AAA size NiMH rechargeables. **Why not use the highest capacity batteries for the job?**

Routinely found for about $20 including shipping for a 10-pack good for 10 handsets (**Lenmar Pro1010**) or $10 with shipping for a four-pack good for two handsets (**Lenmar Pro408B**). **After two years and NO failures, I am pleased with the Panasonic cordless phones *and* the Lenmar batteries.** The Lenmar battery prices were less than for any others I found in the quantities needed.

CHAPTER 8

REPLACEMENT BATTERIES

FOR CAMERAS, MOBILE AND CORDLESS PHONES, MORE

If your devices and gadgets use rechargeable batteries, this chapter is for you! Cameras, camcorders, mobile and cordless phones, plus laptop computers are the main receptacles of rechargeables. Be smart when purchasing replacements.

My go-to rechargeable battery company is **Lenmar (Lenmar.com)**. I particularly look to them for rechargeable AA and AAA batteries, with among the highest available power ratings and the best prices.

Try them for other rechargeable battery needs, too. Search on their Site for compatible batteries for other devices. If theirs have the same or nearly the same capacity as the original, here is what to do. Find their replacement number for your original device battery and note the Lenmar price on their Website.

Next, copy the Lenmar part number of that replacement battery. Then search online for **Lenmar,** plus the **part number**, to find the best online price for the Lenmar battery required. For example, digital camera replacement batteries with a two–year warranty are at considerable savings from Lenmar. When I search online for the best price, I often find the cost for two, including shipping, is less than for one name brand or original equipment camera battery of similar power rating. I've always had good luck with the brand and have recommended them for several years.

CHAPTER 9

VACUUMS, ETC.

VACUUMS AND HARD FLOOR CARE – CLEAR WINNERS

One would think that choosing and then buying a vacuum cleaner should be an easy task, but, like so many other things, it is not.

We consumers are subjected to marketing hype without being provided the important information, the information that genuinely helps in making an informed decision. I like to cut through the clutter, of course, but only after my own hands-on evaluation. Marketing hype cannot replace hands-on experience, knowing what one is looking for and looking at. Unfortunately, consumers are generally not afforded this opportunity to use and compare, nor are most interested in taking the time to do so. Mr. Gadget®, at your service!

If you have hard floors and little to no carpeting, a traditional vacuum may be overkill. Vacuums are all designed with a rotating "beater bar" also called a "brush bar," that along with suction is designed to rotate and draw dirt from the pile of a carpet. However, some users may want and benefit from a traditional vacuum, especially if the rotating brush can be switched off. The hard floor care info is coming later, so hold on, please. Back to vacuum cleaners . . .

Despite what makers of bag-type vacuum cleaners may tell you about bagless vacs, some bagless vacuums are considerably better than their bagged counterparts. The issue claimed by bagged vac makers is that there is much allergen-infused dust strewn about when emptying a bagless bin. Yes, there can be, but I have found it easy to be careful and avoid the mess by placing the dirt bin deep in a new and empty 13-gallon garbage bag that is inside a 13-gallon typical kitchen-size garbage can and then emptying. Do this indoors or out, as you find best. End of concern.

Bag-type vacuums, by design and technology, lose suction as the *inside* of the bag becomes impacted with dirt. That is the way a bagged vacuum works. The bag is porous, but only allows air to escape, which is why the bag expands with air as it is used. I've found the bags to be more of a nuisance (and added expense) than dealing with a bagless dirt bin. Some bagless models are difficult to maintain, including keeping filters clean and free of debris. Some are also difficult to clear of clogs that can occur in even the best of the best. Some are difficult to disassemble to the point needed in order to clean, say, hair from the beater bar brush. All of this has been taken into consideration in my recommendation.

The best of the bagless lot by a wide margin is Dyson (dyson.com), and, in particular, the DC28 Animal with Airmuscle (bit.ly/uSGazA - link shortened for your convenience). There are other Dyson Animal models but only the DC28 has Airmuscle technology. Why Dyson and why this model?

Dyson vacuums, for the most part and in most models, are the easiest to operate and maintain as they are the most intelligently designed and engineered. They may cost more than ordinary vacuums, but they also save money for the consumer throughout ownership. Not only are there no bags to change, ever, but it is also rare that a Dyson will require a replacement belt, if the particular Dyson model is equipped with a drive belt. The DC28 has no drive belt. Its drive is electric. ALL vacuums with a rotating brush require attention in that, depending upon your environment, pet or human hair, or other stringy material may need to be removed (I call this giving your vacuum a haircut!) as often as after every use. This often overlooked and critically important area is the source of much dissatisfaction with *any* vacuum cleaner. I mention this because Dyson vacs have the distinction of being engineered with critical areas' disassembly for maintenance in mind, and this includes the brush head lower plate, which must be removed in order to reveal the brush itself for this haircut procedure. On my Dyson, all it takes is a coin to quarter-turn plastic screws in this area. Remove the plate, use scissors to cut the hair or other wrapped-around material, and then re-fasten the plate. Videos for this and other maintenance items are on the Dyson Website under the Support link. NO other vacuum is as easy to maintain.

Dyson advertises that their vacuums do not lose suction. It's true! Others do, even other bagless models claiming they do not. This is because only Dyson uses their proprietary and patented Root Cyclone technology. Yes, others claim their own cyclonic technology, but others are not designed the same and certainly not as well. Others can clog more easily than Dyson. Others have filters in need of constant attention, as frequently as after every use. The others' filters clog with dirt and negatively affect suction. Dyson filters are both permanent and low maintenance, requiring usually only periodic rinsing, 24-hour air-drying and then putting back in place. Dyson's filters are not in the airflow in a position to become clogged, as are others'.

ALL Dysons do the best vacuuming job, in my view, in their class and in their price range. The DC28 with Airmuscle is unique in the line with its motorized capability to adjust head height to fit bare floor with the brush automatically switched off, low, medium and deep pile shag carpet types. All other Dysons have a floating head that either manually or automatically adjusts to the proper height. The distinction with the DC28 is that the motorized head HOLDS it in place better and makes it able to create a greater "seal" between the head and the surface below. It is this seal that allows for greater suction and deeper cleaning down to the carpet's base, a feat I personally witnessed in just one other vacuum, a **$2400 Kirby** (**kirby.com**), sold door-to-door only.

At a retail price of about $650, this Dyson is a far greater value with no bags, no belt and likely no need for periodic service if well maintained by the owner. It's easy! Just keep emptying the crud from the bin, the airway if needed, the brush and, perhaps annually, remove the bin, separate its parts (they snap apart and back together) and give it a thorough tapping to remove any accumulated household dust from the cyclonic chamber. Every three months, clean the two filters; one is under the dirt bin and one is immediately to the side of it, with both locations well marked and visible. Under the *Support* link at the Dyson site there is a video to demonstrate the simplicity of this procedure.

Sorry, other brands, but you do not compete. Yes, there are less expensive brands and models, but they all require more effort and cannot clean as well as this Dyson or less expensive Dysons. They all will undoubtedly require an expense to maintain, as well. Even other,

less expensive Dyson models, those without this unique deep cleaning capability, clean better than other models costing less as well as others costing more. And ALL others are much more complicated and costly to maintain than any Dyson model. I think of cost as not just raw dollars, but also inconvenience and attention that is needed with others that is NOT needed by owners of Dyson vacuums.

Buy a Dyson and keep it for many, many years, well beyond the likely point of displeasure and expensive maintenance with other brands.

What about hard floors? Steam cleaners do the best and most economical job on hard *sealed* floors. Stone or other surfaces that are not sealed should not be cleaned with steam or anything else without following the recommendation of the surface's maker or installer. Follow their guidelines.

Sealed floors do very nicely with regular steam cleaning. Of course, the floor must first be cleaned of any staining or caked-on dirt. If your floors are devoid of carpeting or have little of it, you may wish to get only an inexpensive Dyson or even a lesser brand, since the bulk of its use will be to pick up from the surface of a hard floor surface, not needing the brush action. If you can get by with a broom and dustpan, go for it!

Steam cleaners work best on floors that are new or otherwise well maintained from the start. Surprise! Not all steam cleaners are alike. Some have "warm" steam, which does not clean well and certainly does not sanitize as does steam at about 200°F or more, and preferably as close to the magic HOT 212°F point as possible. Read the "candidate's" literature to find the degree of steam the candidate emits, if stated.

Steam is a terrific cleaner AND sanitizer. High temperature steam both cleans and kills germs. Just plain water is the steam's source. I DO NOT recommend any hard floor sanitizer that is not steam only. All those using chemicals are not effective AND are annoying to use, as the chemical has to constantly be pumped out to do the job. Chemicals are bad and expensive AND they can introduce their own odor. Steam is just plain tap water, heated to the point of becoming a gas. Steam cleans!

Remember to be sure the floor is swept or otherwise cleaned of any loose dirt or debris. Steamers are NOT vacuums nor do they remove dust. The floor steamer that I recommend removes ground-in dirt and stuck-on gunk, but occasionally removal may require a spray of plain water and a microfiber cloth (my favorite is **E-Cloth – ecloth.com**). This is not the norm, however. The need for chemical cleaners is rare; try to use only steam.

My pick for the best hard floor sanitizing steam cleaner? Haan FS20 Plus, about $80 with careful online shopping (haanusa.com). The FS20 runs for about 20 minutes per tankful of water. Purchase includes TWO washable/reusable microfiber pads and a sanitizing tray attachment for refreshing carpeted areas with steam. Despite its shorter than ideal 20-foot cord, the Haan is a standout in performance and ease of use for the price. I've loved mine for more than three years.

Here's the simple drill. Fill the Haan reservoir with tap water and attach a clean microfiber pad using built-in hook & loop attachment points. Switch it on and place the Haan on its supplied foam "staging" pad, and in just a couple of minutes, steam will start to emanate from the 15 jets on the underside of the device. Now, place it on the debris-free sealed hard surface and do nothing more than use a back and forth motion to cover the floor, slowly enough for the steam to do its work. It's a trial and error system, but *so* easy to get it just right. If the pad becomes super-saturated and/or gross with accumulated dirtiness, put on a fresh one and move on. If the sealed hard area is large enough, you might need additional pads. I have not worn any of them out in over three years of regular use.

When the reservoir is exhausted (indicator turns red), and if more cleaning is required, refill, wait about 30 seconds and switch on to go again.

I also use my Haan to clean and sanitize counter tops (yes, a bit awkward, but effective) and the tile floors inside our showers.

Read my reviews of the Dyson DC28 and Haan FS20 at MrGadget.com.

CHAPTER 10

HOUSEHOLD KNIVES

HOUSEHOLD KNIVES – CHOOSE THE RIGHT TOOLS

This is about knives used in the kitchen for food prep and those used at the meal table to cut your food. Here I will cover the knives themselves. The next chapter will present the key to perfect, super-sharp knives, *your* knives that you can and will sharpen, expertly, easily, quickly and effortlessly. You will NEVER need nor want to pay someone else to maintain your knives. First, though, let's start with the knives themselves.

It is not necessary to accumulate several different kitchen knives for most daily needs. And it is not necessary to spend a fortune for knives that will absolutely last beyond your lifetime, to be handed down to a lucky recipient after you've expired, and so on. These make excellent gifts for newlyweds and for singles off on their first adventure away from the nest.

Of course, you may develop a love of knives as I have. And then you may want to get more than the basics. I *do* like to think about which in my collection I want to grab for a particular task. I'll cover the basics and beyond.

There are many steel formulae used in modern knives. I am not such a purist to go there, however! My advice is to look for high carbon stainless steel and let it go at that. The brands I use and recommend are of this type of stainless steel and do not rust in normal use.

Do not buy into the outlandish claim that a knife will stay sharp forever and never need sharpening. This is not possible and often is part of the price paid for inferior and overpriced knives. The catch is that these products' makers will often offer free sharpening for life, if needed, of course, and often for a shipping and handling fee. Don't do it. It's a scam, often part of a door-to-door sales practice, as well as an inconvenience. Follow my advice for a better, sharper and less expensive result without the inconvenience.

These are the basics; only one serrated knife, that being a bread knife (optional, for bread lovers); a paring knife for cutting small, perhaps delicate fruits and veggies – I prefer about a four to six inch blade, though some like shorter lengths of, say, three and one-half inches; a 10-inch chef's knife for cutting or trimming big hunks of meat or other large food objects; a seven-inch Japanese-style Santoku, easily a daily favorite for chopping and slicing just about anything; a 10-inch carving knife, the long, thin one for slicing roasts, hams and carving turkey and other fowl; if you prepare whole fish and want to bone or fillet it, a seven-inch flexible boning/fillet knife. That's about it.

I do not generally recommend buying knives in sets, except, possibly, steak knives (also not serrated, please). Even more basic than the above recommended pieces, perhaps just a couple of knives to start would be a paring and Santoku. Many users would find these to be sufficient. It's always easy to add as the need or mood strikes.

Today, as I check Amazon.com prices for the **Mundial**-brand knives (more on this brand later) in sizes and types specified above, prices are, in order, $35, $22 - $29, $40 - $50, $40 - $50, $41, $34. That's under $250 for a lifetime *plus* of kitchen magic. If just for the two basics above, that's only $29 plus $41. Why not just go for it? That's less than $75.

I recommend using a knife block to accommodate several kitchen knives OR get plastic blade guards to protect each one if they will reside in a drawer. If in a block, and blades are stored up and down as opposed to on their side, rest the top, the thick, non-sharpened upper part facing *down*, with the sharp edge facing up, so the sharp edge is not resting against the bottom of the slot. In this way, the sharp part is not resting against anything so the sharpened edge is not going to be dragged in and out of the block. Steak knives are most often kept in a fitted presentation box or block with fitted slots for each in the set of usually four, six or eight. Empty knife blocks are priced from about $25 - $50. Plastic knife blade guards will run an average of about $5 each.

A very good knife brand you've probably never heard of is **Mundial** (**mundialusa.com**), a Brazilian maker using steel as good as or nearly as good as most from Germany or Japan. I know this from personal experience over several years of using the brand. I've compared their ability to be sharpened to the fine edge I like and to stay sharp against all my other, more expensive knives. The result is about the same. Only the *feel* and *balance* is different, but not so different, in my view, to justify the added cost of more expensive knives *for most users*.

I like a kitchen knife that feels substantial, but not heavy. A knife that is too lightweight feels flimsy and of inferior quality. Also, a knife blade that is flexible, other than the designed-in flexibility of a fillet knife, is NOT desirable and indicates a knife of inferior quality. Other than the flexibility issue, this is very much a personal determination.

Mundial's 5100 Series in black, white, red or pink is, I think, the best *value* in quality kitchen knives. As you learned from their prices listed above, they are not expensive in consideration of their heirloom quality.

If you want to independently learn more about knife blade material types, investigate the difference between forged and stamped knives. You will learn that stamped knives are cheaper and easier to manufacture. The kitchen knives I like best for their feel and quality appearance are forged, not stamped. They cost a bit to a bunch more, but with Mundial knives so reasonably priced, I think it worth the extra cost for their forged knives.

Let us not forget the quality of German knives, regarded as among the world's finest. (Mundial began as a German company more than 80 years ago, and moved its operation to Brazil.) The German brand of excellent quality and best overall value is **Wusthof** (**wusthof.com**). I like to keep things simple so I will not suggest any of the other excellent brands, but shop if you must for other makers' products. The Wusthof knives recommended come with a new and superior finished and sharpened edge (more on this later), while, to date, Mundial is not as careful or as precise with their finished product. However, as you will learn, this is NOT as important as you might have thought (if you ever thought of such a thing at all). Why? Because in the next chapter you are going to learn how easy it is for anyone to properly, expertly sharpen knives at home in mere minutes. Therefore, it matters little how they come from the factory.

The only wild card is choosing steak knives. Unfortunately, Mundial only produces steak knives that are serrated. Here is where there is good value to be found in stamped knives and in other brands of forged cutlery. Remember, stay away from serrated steak knives! Be patient. I will soon explain why.

A good brand of non-serrated stamped stainless steel steak knives to look into is **Chicago Cutlery** (**chicagocutlery.com**). Find sets of four or more for as little as about $25. I've just found a seven-piece set of forged Chicago Cutlery "Cayenne" steak knives on Amazon.com and for only $35, including a wooden storage block. Shop wisely!

Find this type of knife in solid stainless steel and with handles made of synthetic materials as well as natural wood. The most durable will be solid stainless or with handles made from synthetic materials.

Alternately, a good set of six Wusthof steak knives will set you back about $200 - $300, depending upon which series is selected. They are relatively expensive, but to many users they will be worth the investment as you and your guests effortlessly glide though otherwise tough-to-cut meats using these beautifully crafted utensils. They look and feel wonderful, but many of you will not want the added expense of this prestige brand name. I get that! By the way, I recommend using steak knives only when called for. That is, when sharpness counts. In doing so, the knives will require maintenance less often. If you're on a tight budget or if you want to try a "starter" set of steak knives, take my advice above for substantial-feeling, less expensive knives stamped or forged stainless steel.

There are also many Japanese-made knives of distinction. Why not highlight one or more Japanese brands? The *only* reason has to do with overall value and type. While there is much to offer here, it becomes confusing to most consumers that there are so many different styles and shapes of Japanese cutlery. This is a real knife aficionado's playground. You are encouraged to delve into this new world to at least see what are clearly the most beautiful and unusual knives available, and among the best and most expensive. The most common exception is the Japanese Santoku-style that is a true crossover now popular in most non-Japanese makers' lines and recommended here.

NEVER wash sharp knives in the dishwasher where they may contact objects that can dull them. In addition, some handle materials do not respond well to being dishwasher washed, so observe the manufacturer's guidelines. Hand wash most easily by squeezing the blade between a folded over sponge, with the blade NOT aimed into the fold. Wipe the soapy sponge from handle to tip until it's clean. Then rinse and hand dry, and put away. Do not toss them in to the sink with dishes or other knives. Care will pay dividends; knives will not become dulled by contact with other items nor will the edges become "dinged" from banging into objects that will do damage.

You've probably been wondering why I advise against serrated edge knives for all but a bread knife. Think about it. Have you ever seen a serrated *razor*? There are no serrated *surgical knives*. A serrated edge relies upon high and low points along the edge to assist in gripping and tearing, as well as cutting the object. Serrated knives cannot hold a fine, sharp edge as can a good non-serrated edge. Of equal importance is that a serrated edge is not nearly as easy to sharpen as is one that is not serrated, as you will learn. Bread is the right stuff to cut with a serrated blade, especially hard-crusted artisan breads.

With the exception of the bread knife, in a class by itself, all dual-bevel straight edge blades can be categorized by the angle at which they are sharpened. By dual-bevel, I mean knives that are sharpened to a same angle on both sides of the V-shaped blade. To you "snobs," I will acknowledge the existence of specialty knives sharpened on only one side. There are also other special characteristics of some knife types, but to the normal, everyday consumer like most of us, the V-sharpened basic knife is all we ever need or use.

Here is the meat of the knife story: *It's all in the angle.*

Within this category are so-called American and European angled blades, sharpened at *about* 20-degrees per side of the "V." Many of the Japanese-supplied good quality knives of this type are sharpened to *about* 15-degrees per side. Keep these figures in mind. I'll uncomplicate it soon.

ALWAYS use a cutting board. ALWAYS. No exceptions! I prefer ones made of polypropylene, that slightly soft non-porous plastic material, or a sealed hard wood or bamboo. The advantage to this type of plastic is that it is easily cleaned and will not absorb odors, juices, or bacteria. Of course, there is one exception to this rule – steak knives at the dinner table. Meat is cut against the plate underneath. They'll require more occasional sharpening, but fret not, that info is coming soon.

Beyond the basic knife type and size recommendations mentioned above is an example for the enthusiast, the kitchen chef who loves the hobby of food prep. I have recently begun to use a Japanese-style *Nakiri* knife from Wusthof, their Classic model 4191. This distinct, nearly seven-inch long implement appears to be almost rectangular, with only a slight rounding at the blade's leading edge. It is tall, about one and three-quarters of an inch, and with a blunt, tall front. That's right, NO point. Yet, this handy *very* thin-bladed knife is my new favorite cleaver for veggie slicing, dicing and chopping. The same knife with a hollow, "vented" blade so foods are less apt to stick is their model 4193. The cost? Only about $100. Search the word "nakiri" online to see this unusual-shaped knife style from this and other makers. It's not for everyone, but a good example of a Japanese-style specialty knife gaining in popularity that could become a mainstay in the kitchen for many of you.

What about ceramic knives? If you have not seen them, you are likely to see them soon. Ceramic knives are not new. This type of blade material gained a following as premium goods commanding a premium price from Japanese maker Kyocera (search *Kyocera ceramic knives*). Ceramic knives have a blade, as the name says, made from ceramic material and not metal. The process yields an extremely lightweight blade, usually stark white in color. Their appeal is said to be their high abilities for sharpness and lack of imparting a metallic "taste" as other knives can be said to do, though I have never encountered a metallic taste from any of my stainless steel knives. Ceramic knives are more delicate in all regards than are their metal brethren. Ideally used for slicing and gentle chopping of veggies and other soft foods, ceramic knives are NOT to be used to slice, for example, meats with bone in or other hard foods. Hitting bone can spell disaster for a ceramic knife, as chipping becomes a distinct possibility.

In the last two years, inexpensive Chinese-made ceramic knives have been introduced with low, low prices and of less than the highest quality. Still, ALL ceramic knives, be they high or lower quality, suffer from the same somewhat delicate nature. My take? For most users, including your humble reporter, I'm not making the switch. Try one if you must. My sharpening method is SO easy and fool proof, however, and my knives are SO sharp that it is not worth having such delicate tools.

And what about re-sharpening, as all knives will most certainly require in a few weeks to months after first use? Traditional and modern sharpening methods for steel knives are not the same as for ceramic knives. Why bother with yet another method and the added expense of a sharpener uniquely for the special needs of ceramic knives? Stick with good ol' steel knives.

One arena in which ceramic sharpness is of benefit is in kitchen tools, principally peelers. Get a ceramic-bladed peeler (search *ceramic vegetable peeler*). These tools are not expensive and do a great job. Leave the other knife-work in the kitchen to traditional steel knives!

Now that you know the basics of *what* to buy, it's time to think about what comes next. That is, care and maintenance. You know they won't stay sharp forever on their own, regardless of forever sharp claims by some knife makers!

Some knives are delivered to the end user (that's you) perfectly sharpened. Some are not so well sharpened and will require immediate attention to make them into the superb tool they will become after your care.

And that brings us to the next chapter, on the topic of *knife sharpening*.

STEVE KRUSCHEN

CHAPTER 11

KNIFE SHARPENING 101

YES, *YOU* CAN DO IT!

I am confident that most readers have no idea and, to now, likely no interest in this topic. However, I ask that you read what is on the topic below. It can be a life changer! I hope each of you will come away with knowledge AND the desire to have über sharp knives for all their advantages and that you learn to enjoy using and maintaining sharp knives, saving money and time. And mostly, I hope you pass along this information to others. It all seems so obvious to me and brings such joy to others.

Even as a Cub Scout and then as a Boy Scout I had a fascination with knives. Moreover, I was interested in *sharp* knives. It was as a Scout that I learned to sharpen knives *by hand* using special stones designed for the job. I became quite proficient at it, too! My Scout knives were always very sharp. I transferred this skill to all our kitchen knives and the appreciation for sharp knives has stuck with me ever since.

As a young boy, I also developed an interest in cooking. Cooking and knife use, as well as eating, all go together, don't they?

I shall never forget being taught that, "**A sharp knife is a safe knife.**" On the surface, this makes no sense. That is, until I accidentally cut myself with a not very sharp knife and again, by accident, of course, with one that was quite sharp, *shaving* sharp.

A dull knife can create a jagged wound, one that is nasty to behold and that may not heal without an unsightly scar. And I recall that those dullish knife wounds also were the most painful. Similarly, I know from experience that a very sharp knife wound may hardly be felt initially and its appearance is clean looking, except for the blood. The sharp knife wounds invariably have healed with nary a trace.

This is not *the* reason your knives are better off sharp, however. Using less than sharpest knives is dangerous, too, because they more easily slip off target and off course, and there goes another cut to tend and mend! Alternatively, the really sharp knife, in the kitchen or at the meal table, is easily maneuvered with precision. Whether thin-slicing a tomato (a demo that always gets high marks!), slicing or trimming fat from meats, filleting fish or cutting other delectables, a sharp knife becomes a pleasure to use. Would you find it amazing or even interesting to learn that one never should need to exert much pressure while using a knife, as if sawing through wood? A sharp knife assures that cutting and slicing through *most* foods requires little downward pressure. **The knife does ALL the work. The knife operator's strength is NOT a factor when cutting through a slab of meat!**

If you've never had the opportunity to become accustomed to, let alone to only occasionally use a fabulously sharp knife, you may think I am not all there above the shoulders. And if you're not interested in kitchen duties, you probably don't care and have no interest in this topic, which is too bad, really. There is a joy that comes to those who have experienced outstanding knife sharpness and who appreciate using them. I've witnessed this so many times with family and friends I've introduced to what *you* can also achieve so easily. If only to experience sharp steak knives I hope you will have such an opportunity and can appreciate what is occurring.

May I simply ask that you take a leap of faith here, at least just for now? Please continue.

Are your knives sharp, as sharp as they were when new (if they were sharp then), or even sharper? What do you do, if anything, to keep them sharp? If you send them out or take them to a professional for sharpening, what is the financial cost and the impact upon your time? Or, do you not bother because of the inconvenience or cost? Until you learn what is to follow, I admit that it can be an expensive and time-consuming exercise to either manually sharpen or to have knives "done" elsewhere. I would bet that most of you either don't think about it OR don't bother due to cost or inconvenience.

I hope to change this for the better for everyone by my advice. I want each of you to love your knives and to respect them for the fun they can create in your food prep and in eating your meal. I also want to encourage each of you to use safe practices when "operating" your knives, whether dull or sharp. Search YouTube for the many instructional videos on safe kitchen knife use. Search **how to** and then add such things as **use a kitchen knife** or **use a Santoku knife** and so on. You'll learn techniques that will make it more fun to be in the kitchen.

In keeping with my personal objective of also simplifying what some find complicated, in short, **knife sharpening is SO easy**! In the late 1980s, I learned about a sharpening method/system that was to forever change how I approached and appreciated the topic. The company is **EdgeCraft** and the products are under their **Chef'sChoice** brand (**chefschoice.com**).

Let me just say that I have long ago given up sharpening by hand using stones and other stationary surfaces *forever*. I NEVER use a sharpening rod, either. These do not really *sharpen* hard stainless steel knives. When you see a professional chef or butcher using a steel rod, known as a *sharpening steel*, the knife against which the steel is being used is, invariably, NOT a super-hard stainless blade of the type consumers use. Their carving knives are made from softer steel, of the type designed to be quickly and easily sharpened again and again, and then retired and **replaced** with regularity. Many professional chef's or carving knives are stamped, not forged. Please take my advice and just forget about this tool. It is not needed and cannot help you to achieve long-lasting or precision sharpness of your consumer knives. I know this sounds like heresy!

Sharpening is all about precision, including the angle at which the knife is sharpened. We humans simply cannot hold a knife at a precise, measured angle, whether against a steel rod or while honing against a stone. Just *fahgetaboutit!*

What about the many sharpening tools that consist of a V-shaped metal trench through which a knife is drawn during the sharpening activity? Sadly, ALL of these devices rely on scraping and tearing the knife's steel edge as it traverses the V-groove. It is also impossible to cleanly, precisely maintain the desired angle. Those devices mimic a sharp edge by creating micro-jagged ridges along the blade. The edge is no longer smooth as it was designed and intended. When testing the knife's sharpness after using such a device, the knife's edge merely tears, not cuts the paper, meat, tomato or whatever is used to prove sharpness. In my view, these products ruin the fine edge. They do not "sharpen." I would never use such a tool again, having tried many of them.

There are also some "V" trench devices made of ceramic rods. Ceramics are the right material for some finish sharpening duties, but the effectiveness of ceramics in sharpening a dull knife is slow and cumbersome, and unnecessary. There are better and inexpensive methods and materials that yield excellent results in a fraction of the time. Please, don't bother. Keep reading for my recommendations. I've done all the legwork, all the testing so you can benefit from both lowest cost alternatives as well as my overall and ultimate solutions that are much less costly and less time intensive than ANY other way that can yield the objective of shaving sharpness. Trust me!

I still try other products purporting to be the *be all* and *end all* sharpening tool, but only one line reigns supreme – **Chef'sChoice**. If that should change, I'll let you know. Until then, **stick with the advice that follows**.

Chef'sChoice produces *manual* and *electric* sharpeners. Their precision comes from two factors. The sharpening is accomplished using crushed diamonds that are bonded to a rotating wheel. The diamonds, harder than any steel, safely remove tiny bits of metal during the sharpening operation. The second piece of the puzzle comes from the precise angles at which Chef'sChoice sharpeners hold the blades while being sharpened. No unguided hand can hold a blade with such precision while sharpening, whether at 20 degrees or 15 degrees per side of the blade. It is this precision sharpening angle that makes all the difference in the world, as I learned.

Having tested various Chef'sChoice models since 1989, here is what I have learned and what I practice here at *Gadget Central*. Until a couple of years ago, all my knives were super sharp using the Chef'sChoice model 120. This electric device performs the sharpening operation at the customary American/European approximate 20-degree angle per side. This is perfectly fine, since the knives sharpened were produced at *about* that angle.

A quiet *evolution* is underway. Some knife suppliers are shipping their wares at the more precise and ultimately sharper Japanese-inspired standard of about 15 degrees per side. I was alerted to this change, playing catch up to the newer Japanese precision, first from **Wusthof**, on some of their products (**wusthof.com**). The company calls this PEtec, or Precision Edge technology, which simply put, uses computers and special machines to achieve this more precise and sharper edge of about 14-degrees per side. Now, Wusthof has changed its entire line to PEtec sharp! Surely, other knife suppliers will follow in this change. What is the consumer to do to maintain this *new sharpness*? Ah, this is the concern, precisely!!

Chef'sChoice to the rescue! Their model XV15 provides a precision triple-bevel edge at this Japanese-style and inspired angle. The long name of the sharpener is **Chef'sChoice® Trizor® XV™ Sharpener EdgeSelect® #15**, but it is easily found online by the **XV15** label. My hands-on, real-world use tests revealed that my knives, ALL my knives, responded perfectly to being converted and sharpened to this new, narrower, and ultimately sharper angle. I experienced superior cutting performance with *no* loss of longevity of the sharpened edge, including my steak knives! My Wusthof knives that began life in this new "PEtec" angle are now easily maintained at about the same angle. All my knives are now either converted to or maintained at this angle. **This is what I recommend to each of you. It is SO simple**.

Some knives are honed to even greater sharpness, to an even narrower angle. One example is my prized Wusthof Nakiri knife mentioned in the previous chapter, delivered to consumers sharpened to an incredible 11-degrees per side. Now what? Eleven-degree sharpeners do not exist, at least none I could find. How can that knife be maintained as new at this narrower angle? It can't! I am NOT worried about it and had I not asked, I would not have even known that it is at this even narrower angle. My anecdotal testing and actual sharpening reveal that there is not a perceptible difference between the super-fine original 11-degree angle and a Chef'sChoice created and maintained 15-degree edge on this Nakiri knife. There IS an easily perceived difference between 20-degree edges and 15-degrees, however. The Nakiri holds its edge, though! I have no reason to believe it will not live out its days in my hands with its Chef'sChoice 15-degree sharpness and do just fine.

I've proved time and again how easy it is to do a great job with Chef'sChoice by sharpening friends' knives. In each case, without exception, the reaction is the same. Friends are amazed and overwhelmingly pleased with the result. I did it again just recently with a new friend's family knives. The family has found new love for their old and amazingly dull knives, never as sharp as when I returned them after the **Chef'sChoice XV15** treatment.

If it is a goal to sharpen one or more thick knives that are currently butter-knife dull, this is the wrong tool! So much metal must be removed to reveal the sharp edge lying somewhere beneath that other tools are better suited to get such knives into shape so they can be sharpened by an XV15. This sharpener and others recommended here are not designed to put an edge on a dull knife as described. You'll be money and time ahead if you just pay a pro to get such a knife in basic sharp shape, and then use Chef'sChoice to finish the job and maintain the edge.

For about $150 from many online retailers you can have this same stunning electric tool that will, at once, convert your non-serrated knives to this new and super-sharp angle, and is sure to last through as much as 20-30 years of normal household use. This is my own *estimation* of product longevity based upon my experience with Chef'sChoice sharpeners. The Chef'sChoice warranty is three years on this product. I am confident, however, that consumers will enjoy many, many years of perfect service from this product. Such is my experience with ALL Chef'sChoice electric devices (and their own Chef'sChoice-brand knives, too).

If you are undecided and want some knives at 15-degrees and some at 20-degrees, they've got you covered with the **Chef'sChoice Model 1520**, about $170 from online retailers. This model features a two-sided slot for 20-degree knives and one for 15-degree knives, plus a third, a finishing and polishing slot for both.

Using these Chef'sChoice electric sharpeners is as easy as it gets, with no guesswork. In three-slotted, single angle sharpeners (like the XV15), I generally start with slot one taking a few slow and steady swipes per side of each knife, then move to slot two, doing the same, and then finishing with slot three. This puts a complete fresh and new edge on the knife. Moving forward, occasional swipes through the final slot may be all that is required to maintain optimal sharpness. Otherwise, it has been my experience that slot two and then the third stage does the job. It is rare to need to do it all again through all three stages, but it is just so easy to do! Regardless of circumstances, a fine and finished edge is just minutes away.

Each slot guides knives at the precise angle taking all guesswork out of the process. Perfection every time is what you will experience. The procedure is truly mistake-proof.

For those of you not inclined to convert and use knives at this *new sharp* (and I cannot imagine why not), I recommend **Chef'sChoice model 120**. This model is also a three-stage device, and produces precision sharpness at an approximate 20-degree angle per side.

Had I not experienced the excellence and difference between the 15-degree edge and the 20-degree edge, I would not have believed there could be such a demonstrable difference. I strongly urge you to go with this better and sharper-angled edge.

Both models also may be used in their third stages on serrated knives. The third stage will remove minor imperfections in the scalloped serrations. The other stages are NOT to be used on serrated knives as they will not and cannot fix or sharpen within the scallops. I'm afraid there is only one way to properly sharpen the type of serrated edge that can be sharpened, and that is either by returning it to the manufacturer if such service is offered, having the knife professionally sharpened by a local cutlery expert or by using a small and precise crushed diamond coated sharpening rod by hand. This operation can range from a quick fix to a labor-intensive, time-consuming project. Chef'sChoice markets their own rod-type sharpeners that can correct minor imperfections in serrated edges. Get models **M410** and **M412**.

Also, do not mistake what I call classic scalloped serrated knife-edges for other, more radical designs. If your knife has many and varied finely sharpened points along the edge or anything other than simple rounded scalloped cutouts along the edge, do not attempt to re-sharpen or otherwise maintain that intricate edge. You are bound to fail in your efforts and may injure yourself or others in the process. This is a good reason to stay away from non-standard knife-edge designs.

Chef'sChoice also provides manual draw-through diamond hone sharpeners at lower cost, **great for knives already at the desired angle**. They're also recommended for hunters and campers to take along. Get **Chef'sChoice Model 463** for Asian-style 15-degree knives and **Model 464** for **American/European 20-degree edges**. Both cost about $30 to $40 from Amazon.com. For anyone indecisive about maintaining everything at a single angle, get **Chef'sChoice Model 4623**, at about $30 from Amazon.com, with two diamond hone slots for each angle – one with more aggressive action and one for the finishing touch. These can sharpen most not-too-dull knives to a fine, shaving-sharp edge at the desired angle, and do it better than anything else on the planet! **Only the electric models are assured of sharpening just about ANY dull knife to perfection. Only the electrics can easily create a new edge on a really, really, dull knife.**

If you are still on the fence and cannot decide, read more thoroughly on the Chef'sChoice Website for the particulars on each model, then make a decision. **If you want me to make the decision for you, I'd be delighted –**

convert to or maintain everything at 15-degrees and don't look back. Get the **Model XV15** as the most versatile sharpening tool, even for specialty Japanese knives, angle-sharpened on one side only. If space and budget considerations are paramount, go with the recommended manual sharpeners, though I would not attempt to use a manual sharpener to convert from 20-degrees to 15-degrees. Is that simple enough for you?

Chef'sChoice sharpened and maintained knives are better and sharper than a highly paid professional can produce. As a matter of fact, I know there are so-called professional knife sharpeners who use nothing more than Chef'sChoice sharpeners on clients' cutlery, and they charge a premium for the job. **Make no mistake, you CAN do this easily, and you WILL save time and money. As a bonus, your sharper knives will be safer, as well.**

Off the knife topic, but still on the EdgeCraft/Chef'sChoice topic, the company deserves recognition for their other many and varied products. ALL are of excellent, even superior quality. Please take the opportunity to explore at their Website and know that I am comfortable and proud to recommend ALL of their kitchen appliances, including **Waffle Makers**, **Hot Beverage Appliances** and **Electric Food Slicers**, and also their **Trizor Professional Cutlery** and **M490 Manual Scissors Sharpener**. All are as good as and usually better than anything else in their product categories.

CHAPTER 12

FLASHLIGHTS

HIGH TECH, HIGH PERFORMANCE
RE-THINK YOUR PERCEPTION OF A FLASHLIGHT

I truly love what I am privileged to do.

What kind of a crazy person would be so presumptuous to think that anyone would care what one thinks about flashlights? Moreover, what kind of crazy person is that *into* flashlights in the first place? Guilty on both counts!

If you are already predisposed to a fascination with such things, it is possible that you already know, already have searched for what others find to be the latest and greatest flashlights. In that case, you are already hooked.

I would like to believe that readers on this and other subjects covered in these pages are not already *in the know*. It is hoped that I am reaching those who otherwise would never search for such info. In doing so, it is also an objective to light a fire of interest in at least some of the topics covered here, and that readers would benefit from my research and recommendations. This chapter is a perfect example.

Flashlights? Yes! Many years ago, before the introduction of high-powered LEDs, the range of excellent flashlight choices was quite limited.

My best remembrance, and one that will resonate with many of you is this: I can recall so many times opening the family's kitchen "hardware" drawer and reaching for a flashlight only to discover that the batteries were dead. You, too? Or I'd need the light to be able to find the source of an electrical problem and the light would dim to an unusable level after only a short time. I also need to mention another failing of traditional flashlights – the bulb! The filament inside those old-school bulbs was delicate and short lived even under ideal circumstances. If the flashlight was dropped, it was customary for the bulb to fail instantly! Who needs that today?

Flashlights *were* for emergencies. We rarely used them otherwise. I'm here to tell you that modern technology has made it possible to expand the breadth of opportunities when RELIABLE and brighter, longer-lasting-than-imagined flashlights come in handy.

Having a little, thumb-size and very bright light on one's keychain can be a real lifesaver. A grab 'n go light for nighttime dog walks would be useful, wouldn't it? Wherever there are such natural phenomena as earthquakes, tornadoes, floods, landslides or any other times when it may be dark, even totally, pitch black dark for a few minutes or even days, wouldn't it be wonderful to have conveniently at hand - LIGHT? And I mean enough light by which to function in some semblance of comfort and normalcy?

Unless you have been unexpectedly plunged into the total absence of light it may be difficult to fully comprehend just how much it is that light makes us feel normal, even whole. This is the extreme that, I know, is not something that most of us will have to endure or for which we need to be prepared.

Even in normal, everyday existence, however, flashlights can prove beneficial. A few examples of benefitting from a light at hand include wanting to find what has dropped to the floor in a darkened restaurant, theater or in a car. How about wanting to light the way on a hike or during a camping trip? Do you ever think about exploring the out-of-doors at night during travel to exotic lands? Or visiting some of the most interesting and noteworthy caves? Or walking along dimly lit passages? Or reading a restaurant menu or a theater playbill in the dimmest light? Or simply having a light, just in case of need, while on a walk? Get the idea?

In the past, flashlight illumination was both not so bright AND fleeting, due to inefficiencies of bulbs and batteries. Modern technology has provided the opportunity to re-think the topic, and here I will clue you in to a few favorites among many, many available great products in the category.

Were I the potentate around these parts, I would provide every person a little pocket light for those times of need. Yes, one for every person's pocket or purse would be my plan! The ONE light I'd choose would be a **Photon Micro-Light II** (**laughingrabbitinc.com**), retailing for **about $12** and proudly made in the USA by the originator and leader in this category. **Search for the best online price!** These lights make perfect gifts for anyone.

I use and recommend this particular, surprisingly, even astonishingly bright model because it is activated with a squeeze AND includes a little switch that can keep it on if needed. It is just not practical to continually, continuously squeeze to activate it if needed for more than a few seconds. However, the company makes the more basic, switchless model as well as one called **X-Light Micro, about $10**, with a single, multi-function clickable button/switch that can keep it on and also switch to different modes. There are many other models from which to choose, but you already know my preference and why this is so. The best advice is to visit the manufacturer's link above, determine the type or types best suited for the need, and also including LED colors if other than traditional white, then shop for best pricing.

With intermittent use, the disposable, easily replaceable flat Lithium batteries will last at FULL brightness for many, many years. If left on continuously, figure about 12 hours of use at near full brightness. Compare this to your old flashlight in the drawer with batteries that pooped out in as little as two hours and would also leak in a year or two. The disposable Lithium batteries in recommended Photon lights have a shelf life if unused for at least 10 years. These fine products are great value, easily lasting for years and years and years. They are that good.

For everyday use at home or in the office, to place in the kitchen tool drawer or to place in the car's glove box, I have another suggestion. This one is also what we use here in earthquake country, USA, otherwise known as Southern California. We hang one on a little nail on every inside door's frame because one never knows when Mother Nature is going to get angry, *very* angry. Insert your own natural calamity of choice here. This is also a light I take along on every trip. I like the idea of, get this, 40 hours of FULL BRIGHTNESS in this outstanding, completely waterproof and tested to a depth of 160 feet, *pretty* bright light with simple and foolproof twist on/off operation. It comes with and uses three readily available and inexpensive AA alkaline batteries, too. The light? It's from my friends at the **C. Crane**

Company in Northern California (ccrane.com), called **CC Trek** and selling for **$29.95 each**, or less if purchased in greater numbers.

Here is a little "aside" on these types of "round" AA or AAA alkaline batteries – buy the least expensive ones you can find. The name brands are all nearly identical in performance to the often less expensive store brands. If you see anything indicating or even suggesting super-performance from one alkaline battery or another, ignore the hype. They're all about the same. I buy them in quantity when I find them on sale OR, if you're a Costco member as I am, get their bulk pack of 48 AA Kirkland-brand alkaline cells for about $10! Do not keep these or other batteries in the fridge, either. That's an old wives' tale!

One may, at first glance, think the **CC Trek** light is expensive. For a rugged light that will last as many years as you like, even into the next life, it's a one-time purchase! The LEDs will likely never wear out and the lights are ruggedly built, virtually indestructible. Of greatest importance is that they are ultimately reliable. And don't forget the superior runtime per set of batteries. I am unaware of any flashlights that can match ruggedness and runtime. In an emergency or just for convenience, don't you want these characteristics in a flashlight for general use?

This is THE light I provide with confidence to all the *Gadget* children. I know they have light when needed, whether it has been in their dorms or apartments while in college, or afterwards as they have made their way in the world after moving from home and starting their own families.

I can tell you from experience that it is not the brightest flashlight one wants when it counts, but the most reliable. **CC Trek** lights put out just the right amount of light when it is darkest. In these circumstances, a little light goes a long way. When I have been in total darkness, during a years-ago blackout in New York for example, it was MY little ol' **CC Trek** that saved the day and lit the way for countless New Yorkers I helped while on the streets in otherwise total darkness. Not knowing how long the darkness would persist, I didn't care, as I knew I was good to go for all hours of darkness for nearly four days. I left my light on, beam pointed up, centrally located and resting in the coffee carafe in my hotel room during the night while I slept. Enough light was emitted so I could see everything and not worry about stumbling over unfamiliar objects in the room. I always have an extra set of batteries with me, as well. I've had this light with me on dog walks and traipsing about in the wilderness, so I can attest to it being the right light for these and other circumstances, including needing to see under the hood of a car stuck at the side of the road on a desolate road where there was no other lighting.

All our cars also have one of these in the glove box. FYI, a fresh set of AA alkaline batteries has a shelf life of more than five years. I set a reminder in my online calendar in Gmail to change batteries in all our CC Trek lights at the same time every five years. Consider something similar for your needs (that include reminders to change smoke detector batteries annually).

For those wanting more in a flashlight, perhaps for serious trekking, hiking, or just for superior lighting, I recommend enthusiast flashlights from **Fenix (fenixlightus.com)**. Here, users will find LOADS of light with less concern for battery runtime. The best tradeoff of performance and runtime is in certain Fenix-brand flashlights powered by inexpensive AA and AAA alkaline batteries. I exclude those products using disposable Lithium CR123 and other similar batteries due to the high battery replacement cost. Performance is not so greatly enhanced in those lights, so why incur the additional expense? Alkaline batteries are so inexpensive that I care very little about less runtime when I really need or want much more light.

I know this is so different from the "old" thinking with old-fashioned flashlights. That's the point. With so many new and excellent flashlights available, I want to encourage you toward using these better tools and not just relegating them to the drawer for occasional, emergency use. These are the ultimate in reliability, so grab one as needed, so long as you know what to buy that represents such great value.

My favorite **Fenix light** is their **LD10, now in its 5th revision, about $45** from **Amazon.com**, and featuring brightness up to a substantial 132 Lumens on its highest setting (brighter than just about *any* conventional flashlight). Yet this light is powered by a *single* AA alkaline battery for as much as 34 hours on its lowest, nine Lumen setting and 1.5 hours on its highest, brightest setting, with two intermediate settings in between and, of course, more runtime. This light is tiny, but mighty. Read about this and the other recommended Fenix lights at the Fenix link above. This is the light in its 4th revision that I also take along on all travel. When I want to light up the dark when exploring, this is my go-to light.

A step up in the Fenix line is their **LD20**, now in 5th revision (R5), **about $60** from **Amazon.com**. Similar to the LD10, but twice as long, housing a pair of AA batteries and sporting a maximum 180 Lumen beam. At full brightness, expect about an hour and one quarter of top performance.

Tuning it down to the second lowest of four brightness settings, it is expected to provide about 30 Lumens of light for a respectable 15 hours. More adventurous readers may want this one along over the LD10 R5 on their travels.

These two examples represent what I consider to be among the best consumer-level performers in the field of LED flashlights today. They, and other Fenix lights that use standard AA or AAA batteries, deliver precision-focused light as well as anything around. Most of theirs are not simple on/off lights. As indicated, these and others in the line also feature electronic circuits to squeeze every bit of performance possible out of the installed batteries by regulating power at peak levels until the batteries are no longer capable of doing so, instead of gradually dimming over time. In addition, as LED technology improves, Fenix updates their flashlights with better LEDs from a top supplier called CREE. Multiple electronic modes deliver multiple levels of brightness as well as strobe and S-O-S (Morse Code) flashing on some models. Yes, these are the best of the best for the cost, in my view.

The field is growing! Many manufacturers are working to capture market share. Sadly, there are also many LED flashlights of inferior quality.

I recommend steering clear of the mostly inexpensive, multi-LED models found on the counters of many gas stations and convenience stores. Experience with these has provided disappointment at every turn – poor illumination, inferior runtimes, shoddy workmanship, unreliable performance.

If shopping for bargains in the LED flashlight arena, there is one standout brand of which I am aware – **Dorcy (dorcy.com)**. Theirs are very nice, indeed, and at lower cost, but not with the diminutive size/high brightness/long runtime/waterproof combo punch of Fenix. Overall, Dorcy LED flashlights are not as sophisticated as are Fenix lights, nor with as many innovative features. Theirs and many other similar brands are the basic LED flashlights, with lower cost, made from less expensive and lower performance CREE and others' LEDs, without the electronics that maintain highest brightness and without multiple modes. Dorcy's optics are not as precise, either. In short, users may expect good brightness for the dollar, but without the long runtimes and other benefits the higher-cost Fenix products bring to the party.

High brightness with long runtimes are important factors. When it really matters and when it is unpredictable how long the light might be needed per use, don't you want the longest runtime possible? This is precisely why I place such faith in **Fenix** for performance with good runtime and why the combo of exemplary runtime with acceptable performance of the **CC Trek** lights cannot be overstated.

Other standouts in the specialty area include a pair of lights from **Pelican** (**pelican.com**). One is the well-reviewed official LAPD flashlight, **model 7060**, routinely found through **Amazon.com** for **about $125**, delivered. This is a no-nonsense tactical, rechargeable light putting out an excellent beam of 160 Lumens, plenty bright for most duty. Featuring redundant body and tail cap switches, runtime per charge is about two hours. This is a great daily-use light but not ideal for the average consumer. It is big and brawny, and other lights costing less, even less bright, would be better for most consumers. Still, for anyone needing daily use of a very bright light, this one is hard to beat. Stepping up the line is the **Pelican 8060**, delivering about 180 Lumens. A bit longer and with only a body switch, the 8060 sports up to about eight hours runtime per charge and is routinely found online for a delivered price of **about $140**.

CHAPTER 13

LED LIGHTING

SUPER EFFICIENT, LONGEST LIFE, LOWEST COST PER WATT

While many consumers bemoan the demise of the standard incandescent light bulb, even hoarding the last remaining stocks of some types, the fact is that **Edison's invention is highly inefficient in a few different ways.**

Allow me, please, to take you from Edison to the LED (Light Emitting Diode), arriving ultimately at some money-saving news about a lot of light. It's a very real look at the future of home lighting, starting NOW.

Along the way, I'll give you the facts that others usually omit from the reports you may have seen on this subject. And I hope you will come away with an appreciation and understanding about why you should not fear the future of lighting, maybe even agreeing with me that this change away from incandescent and CFL lighting is going to end up being good for us all.

LED bulbs can be used in homes, offices and industrial settings. They are not yet ideal for all applications, but they are right, right now, for *some* inside areas. Switching to an LED bulb requires more than simply replacing any existing CFL or incandescent bulb. Installing an LED bulb in the wrong setting, the wrong application, is a guarantee that the user will be dissatisfied, disappointed, likely confused and, perhaps even a bit angry.

Like it or not, standard light bulbs, the ones that have been around for over 100 years, are being phased out. I don't like it, and I wish we could continue to have those good old bulbs around, allowing consumers the choice of what to buy, but even I must admit that it is time to allow technology to show us a better way to light our lives. **The reasons the old bulbs have to go are simple; they use too much electricity, they can get extremely hot, and they don't last very long.** The *average* rated life of a standard, *inexpensive* 60W bulb is 1000-hours or less. And the heat they produce translates into measurably higher cooling costs. It will no longer be economical for companies to produce standard incandescent bulbs when the product is phased out by government edict. Not enough can be legally sold to continue to make the venture profitable. And we in the US are not alone. Incandescent bulbs *are* going away.

We have been coerced into switching to Compact Fluorescent Lamps, or CFLs. They use less energy than a standard bulb, only about 13W for an amount of light equivalent to the old 60W bulb. They don't get as hot and they typically will last at least 8-times longer than a standard bulb. But, all is not rosy with CFLs. Many consumers complain about the look of the light from CFLs. CFLs don't start at full brightness. They may appear to flicker, and you may have noticed, over time, CFLs lose their brightness, little by little. Of course, there's also that issue of **toxic Mercury inside every CFL**!

Despite these drawbacks, CFLs aren't necessarily a bad choice, that is, a bad choice today and for the near future. In fact, in *most* instances they are still the best choice since 60W equivalent CFLs offer the least up-front cost (they can be found for under $1 a bulb). But, what if you want to leave a light on all the time? And how about bulbs that are in inaccessible, hard to reach places such as high ceilings and track lighting that you may have to pay someone to change? Is there a better solution?

In some applications, YES. **LED light bulbs *can be* a cost effective, excellent choice.**

But how do you choose an LED bulb? It's not easy, though it should be! Remember when you could visit the store and simply pick up a 60-, 75-, 100- or 150-Watt bulb? That was so easy. Even with CFLs, it is not so easy, as users must not only consider watt equivalencies, but also projected life ratings, physical size and color temperature along with cost. Each characteristic is a component that determines the CFL's cost.

On the plus side, there is no toxic mercury in any LED bulb. They all start instantly at full brightness and they do not flicker. But here's the challenge: **All LED bulbs are not created equal** (and we are only referencing LEDs of the type that replace household CFLs and conventional household incandescent bulbs in this chapter). **Here is where it gets confusing**, and it makes me angry that government standards promote this confusion, which most LED bulb manufacturers use to their advantage! Most LED bulbs on the market claim a 25,000-hour lifespan, **but only when used four hours per day**. Read what's on the package. In 2012, government mandated regulations specify a lifespan rating based upon *three hours* use per day. I'll circle back to this, but keep it in mind.

Most 60-watt equivalent LED bulbs use about 12.5 watts of energy, almost the same as 60W equivalent CFLs. Where is the energy and cost savings?

I've tested many LED bulbs that have lost their brightness just like CFLs, only much faster. **How satisfied would YOU be if you paid even $10-15 for one of these, only to have it dim and fail in just a few weeks or months?** And how interested would you be in buying more, even if there is a claim of improvement in new models? This is what many consumers have faced by being early adopters of this technology.

I think there is deliberately confusing and misleading information coming from most manufacturers of consumer LEDs! OUR government is NOT helping matters with an unrealistic mandatory rating scale, either. It only adds to consumer confusion. What would it mean to you, a consumer just like me, to see a 60W equivalent LED bulb with fine print on the packaging stating approximately the following information:

The bulb in this packaging is rated to last 25,000 hours, based upon three-hours use per day. It is guaranteed for three years (based upon this usage)."

Wouldn't that be exciting, with charts showing this bulb is *projected* to last nearly 23 years, thought *guaranteed* only three years? You might be inclined to pay handsomely for such a bulb, thinking it might last nearly a quarter of a century! Unfortunately, it is a lie, a theoretical *possibility* based upon projections and accelerated testing. Such testing only simulates long-term use and actually tests for a *much* shorter time. And it is OUR government that is mandating the ratings based on three hours use per day. Don't you think manufacturers will do all they can to take advantage of these numbers? Of course they will! Confusing, isn't it?

Who uses bulbs only three hours per day? Do you? Don't you turn on a light when it is *becoming* dark, then turn it off when you retire for the night, when you leave home or when you know you are finished in that room for quite a while? This would mean that some lights would be on for 10 or more hours per day or as *few* as three or four hours daily. Sure, if you come home to a dark house from work at, say 6 PM and go to bed at nine or 10 PM, that would be the approximately three hours daily use, but the three hours usage model is simply not realistic. I say, let's get real! And what about non-work days? Depending upon the season, aren't lights used considerably more than an average of three hours per day? Aren't there some lights that are on from dusk one day until dawn's early light the next day? That's a lot of hours use per day!

There is only one LED light bulb I know and trust that is rated to last 50,000-hours, is guaranteed for FIVE years, and is designed to deliver this performance even if left on 24-hours a day for five full years! This bulb doesn't come from **GE, Sylvania, Philips**, or any other manufacturer you know. **The bulb is the GeoBulb from the C. Crane Company (ccrane.com).** You've probably never heard of the company (unless you are a Mr. Gadget® follower) or the product, and you won't find GeoBulbs in stores, at least not today!

The GeoBulb was designed from the ground-up by the C. Crane Company in California. With efficiency and longevity in mind, the cool-white GeoBulb produces an amount of light that looks similar to a 60W standard bulb. But here's the surprise, **it barely *sips* electricity using only 7W!** That's about half the electricity of those efficient 60W equivalent CFLs (and half the energy of nearly all competitor's 60W equivalent LED bulbs). And, remember, **it's going to last for 50,000-hours, not 25,000 hours.** Try finding a CFL that can come even close. **That's 7W to get nearly the light from the old 60W standard bulb, and with little discernable heat.**

At the end of GeoBulb's designed-in 50,000 hour life cycle, specs call for it to provide about 70% of its *original* brightness, compared to *less than* 50% of a CFLs original brightness after its stated average 12,000-hour rated life. During the life cycle, the gradual decline of a CFL's light output sneaks up on the user, most of whom do not realize just how dim their CFLs have become after just a few thousand hours in use. **CFLs that I have evaluated and used here at *Gadget Central* have delivered considerably less than their** rated light output mere months after installation. GeoBulb is designed to keep going, with much more light right up to its end, years and years into the future.

Sure, the *initial* cost of an LED bulb is considerably higher than a 60W CFL. But, let's compare *operating* costs. At 12-cents per kilowatt-hour a SINGLE standard bulb using 60W of electricity, left on 24-hours a day, will cost about $61 per year to operate. A comparable CFL using 13W of electricity will cost about $13 per year. But the C. Crane GeoBulb energy-*sipping* 7W of power will cost just $7 per year! That's right, **$7 for an entire YEAR, and that's if it is** *NEVER* **switched off!**

And remember, a standard 60W bulb typically is rated for a lifespan of about 1000 hours. A 13W CFL averages 7,500 hours before it fails, getting noticeably, progressively more dim through its life cycle. **The GeoBulb using its 7W is guaranteed for 5-years and rated for 50,000 hours of continuous use! Nobody can match that. No other manufacturer today offers an LED bulb with this much light using so little energy.** Sure, there have been promises from other makers touting what is *almost available*, but GeoBulb is here NOW, and has been for more than two years! Think about this – the GeoBulb was released in 2009, which means that *every* GeoBulb *ever* sold is still under warranty, even if it has been ON 100% of the time, 24/7/365, since purchase, and it may have up to three more years of warranty remaining. Sure, there are important usage guidelines, as with any product, but I am assuming the GeoBulb is used in that "recommended" fashion. If a GeoBulb is used *only* 10-hours daily, its life may span about 14 years. At only three hours use per day on this 50,000-hour rated bulb, not 25,000 hours like most others, *today's* GeoBulb is *projected* to last about . . . 46 years! I think these numbers are ridiculous, don't you?

Let's go back to the simpler equation – 50,000 hours/five-year guarantee, even if left on 24/7/365. That's GeoBulb. No other company matches C. Crane's guarantee or usage model, and darn few are sure enough of their product to rate their LEDs at 50,000 hours and to offer such a long warranty. Look at the packaging!

Where others have failed, I think C. Crane understands what consumers want and need in LED indoor light bulb technology and the company is the most realistic about specifications.

Where should GeoBulb be installed? GeoBulb is best suited for downward facing applications such as recessed and track lighting. For maximum *apparent* brightness, the GeoBulb outputs most of its light from the top of the bulb. But, I wouldn't be surprised if you also like the amount of light when it's in a table or floor lamp.

What is apparent brightness? Since LED bulbs are different from conventional bulbs, it is difficult to compare the brightness numbers, expressed as "lumens" between LEDs and conventional 360° of light-type bulbs. Though LED bulbs generally emit fewer lumens for a 60W equivalent bulb than does a 60W equivalent CFL, because the light comes out differently, so to speak, the LED may appear to be as bright, even brighter than the CFL. This is especially so of the cool white colors of LEDs versus CFLs. In other words, don't get hung up too much on the lumens values. Let your eyes decide, as I did.

Let's examine why the light from nearly all LED bulbs is in a different pattern from a conventional bulb or CFL. You've probably seen the little wire inside a conventional bulb. When electricity is applied, electricity is converted to heat, which causes the wire, called a filament, to glow and emit light. For the record, *most* of the energy used in that conventional bulb is wasted as heat. Light is sent from all points along the wire and in all directions. This is why there is light all around one of these bulbs.

While there is no filament inside a CFL, the gas inside is what glows and sends out light in all directions.

The individual little LEDs that are inside household LED light bulbs are an example of what is called surface mount technology. The LED itself, an *electronic component*, is on a tiny circuit board. A tiny lens atop each LED directs the light outward from its source on that tiny circuit. Light is emitted outward from the tiny point, but it cannot go in all directions, as there is that barrier to light upon which the LED is mounted. It's like a table lamp sitting on a four-legged table. The underside of the table is not well lit from the lamp on top of the table because the table itself is a barrier to the illumination from the lamp's bulb.

Inside a typical household LED bulb, under the frosted exterior, are several *individual* LEDs, each directional in nature. The frosted outer globe making the whole thing look somewhat like a conventional bulb is designed to diffuse the light from its natural relative and directional pinpoints to a more pleasing wide and seamless appearing output. Individual LEDs under that frosted cover simply cannot (at present) deliver all-around illumination, which is why LED *bulbs* are best suited for applications which take advantage of the light coming from the general area near the top of the bulb and not from the middle and lower portions of the bulb. Technology exists to mitigate the directional nature of LEDs as described, though the energy consumed is nearly double that of GeoBulb and the rated LED life is also half the GeoBulb, at *only* 25,000 hours.

GeoBulb looks like a high-tech version of a standard light bulb. Again, though, the best use for GeoBulb is downward facing, taking advantage of the area from which most of the light is emitted. Because an LED bulb can cost considerably more than the standard bulb or CFL it replaces, it should be used where it is inconvenient to change bulbs and in locations where a light is ON more than it is OFF. In this way, users experience the greatest benefits of cost and convenience.

Try this with an incandescent or CFL! You may know from experience that incandescent bulbs in these installations require quite a bit of attention. You also know that CFLs do not do well if left on most or all of the time. Failure rates are quite high. And you know that CFLs do not do well where it is cold, such as in the ceiling in an enclosed porch or in barns or warehouses in cold areas of the country, or even inside large, walk-in freezers. LEDs do not mind the cold in these applications.

What about color? In my experience, I like and prefer the apparent added brightness and the color of the Cool White GeoBulb. But, it isn't the only choice. GeoBulb comes in three color temperatures representing cool white, soft white and warm white. The Cool white is the brightest and most efficient of the three GeoBulb models available, and the one most like daylight sun in color. This is the one that outputs an amount of usable light most closely resembling a 60W standard bulb. When I look at cool white standard bulbs or cool white CFLs after having the GeoBulb at home, light from the others looks dirty and unnatural by comparison.

It may not matter if you choose a GeoBulb in cool white, soft white or warm white. They all appear to emit a similar amount of light to *my eyes*, especially after becoming accustomed to them and to the light they emit. I lined them up to have a look. Although the colors look different from each other with all three lined up as I have done for my own testing, there is a place for each. Whatever is chosen, you will soon appreciate the amount of light and the money being saved.

Many experts recommend choosing just one LED bulb to start. See how you like it. I agree! Some suggest using the bluish-white cool white type in an entry area or laundry room. The soft white may be more pleasing to the eye in living spaces and the warm white is said to be favored in bedrooms where the yellowish glow is similar to old-fashioned incandescent bulbs, best for where we go to sleep and where we wake up. There is science behind color temperatures, you know!

For more than 22 years I've known, trusted and recommended C. Crane as the best catalog retailer, with the best products, and best customer service, too. It's my go-to company for high performance radios and now for high performance LED lights. I NEVER get complaints from recommending the C. Crane Company. **And I am NOT being compensated in any way for this or any review or recommendation.**

I want to impress upon you the need to compare the different manufacturers' LED bulb specifications when you shop, now and in the future. LED bulbs from any other maker today cannot deliver the goods as do GeoBulb LEDs. It's a bold statement, but I am confident it's true, and you know I would NEVER steer you wrong.

The final point I want to make is that there is still a very good reason to use CFLs. Oh, yes! If you need lots of light, way more than a 60W equivalent, get a higher wattage CFL that's a 100W equivalent or more. This is the most economical way to provide much more light than LED technology can economically deliver *today*, but some of those high output CFLs are quite expensive, too.

What's next? As with any electronic products, and remember that LED light bulbs are electronic products, prices come down and efficiency goes up. Don't look for amazing breakthroughs in this technology any time soon, however. **Progress will come, at least for now, in baby steps**. I expect to see the big names make their move with new household LED lighting products in 2012. They have to wait until it becomes economical on a grand scale. Because there are so many advances in efficiencies and cost, it is difficult for the big names to pull the trigger to produce a product that will sell into the tens of millions, only to have it rendered obsolete through technological advances soon after production begins. On a smaller scale, companies like C. Crane are better positioned to deliver changes we want and need; more light, greater reliably, at lower cost. **Be careful with LEDs that are at far less cost than what seems the norm. There is no way the lower cost can translate into greater efficiencies, more light and the exemplary reliability of products including GeoBulb. I've been burned! Now you are informed of the true characteristics of the beast.**

One desirable characteristic on some consumers' wish lists is for dimmable LEDs. This is not as easy as it might appear. Remember, because we are dealing with an electronic product, it is not a simple matter of dimming as if it was an old incandescent bulb by supplying less voltage to produce less light output. Dimming an LED bulb, that is, an LED bulb designed for consumer household lighting, is a technological challenge that adds cost to the product. Look for this feature to be more prevalent when the cost to provide it becomes only a minor cost factor. Now is not that time. Dimmable LEDs available today at or near the cost of non-dimmable LEDs cannot be as reliable as non-dimmable high quality LED bulbs. Compromises must be employed in order to keep costs down. The compromise must be in the components used, and it is reliability in the form of longevity that must suffer. Is the desire for dimmability worth what is likely a significantly higher cost? If so, there are expensive dimmable choices in high-end LED lighting, but I have not tested samples to recommend.

In the meantime, commercial application of LED technology is where the money is for the big names in this end of the lighting business. Let's say I owned a huge company with 200,000 employees, with facilities all over the country and huge buildings needing lighting inside and out, including in parking lots. In these usage models, lights are on more than they are off. Think of, perhaps, a big clothing retailer such as Macy's or a grocery company such as Kroger. How about a huge hospital or state universities? Or malls all over the country? If a company could show me how to save even 30% of current lighting costs, with a bonus of lower cooling costs, and with fewer failures, wouldn't that save me a ton of money, in effect, adding to the bottom line? Sure it would! As someone in charge, I'd want to listen. That would be a ton of business each time a lighting company made a sale.

Today's LED bulbs made for consumer use are sold one at a time. And remember that home LED bulbs are suitable for only few, specific applications today, primarily because of cost as well as design limitations. This is going to change! Mine are in the inside entry area, outside next to the front door and in the front of the house in an enclosed "globe." I also use a GeoBulb to light the workspace around my main computer. Though not recommended for outside use, I've experienced NO failed GeoBulbs in more than two years of use, with the outside bulbs ON an average of about 12 hours daily. Failures have been many among other, lesser LED bulb brands tested here at *Gadget Central*.

I've tried to give you straight talk and useful information about household LED bulbs. I hope you are as excited about this technology as I am. Get more information about **GeoBulb** online at **GeoBulb.com**. **You can also call C. Crane at 1-800-522-8863**. You will be treated with respect by the helpful C. Crane staff located in Northern California. They love to answer questions. You will never experience high-pressure sales tactics, either. I know you'll be impressed.

CHAPTER 14

PORTABLE POWER

BATTERY-IN-A-BOX – EMERGENCY POWER, AIR FILL, JUMP-START

The **Energizer**-branded **All-In-One, Model 84020, (bit.ly/tgPuLM** – link shortened for your convenience) **about $138, is, far and away, the best combo battery booster/jump starter, small air compressor, AC and DC power supply of its kind I've ever seen**. And I've seen several of them.

This accessory is worthy of being resident in most homes across the country. Wherever there is the possibility of a power outage due to hurricane, earthquake or tornado, Energizer All-In-One will be appreciated. Whenever one may need a quick jump-start, it's handy to have an Energizer All-In-One. Whenever one may need to inflate an auto, motorcycle or bicycle tire, or a ball, or a piece of exercise equipment, a raft or inner tube for water fun, Energizer All-In-One is the right tool, at the ready.

I've tried other brands, many of which are similar in capability and basic functionality, but no other has the smarts found in this one. I'll address this shortly.

Even if you have AAA or some other service that can be summoned to the rescue if your car battery is dead (at home), why bother calling and waiting when you can so easily jump start your car or light truck with one of these?

Are your volleyballs, basketballs, footballs, beach balls and that giant exercise ball properly inflated? How about your car, motorcycle, truck, trailer, or spare tire? Need to top off the air in any of these? You don't have to drive to find a gas station or other service facility to do this once you have one of these handy gadgets. The compressor and gauge are capable of delivering up to 250 PSI, surely more than any need dictates. It can fill a typical auto tire in about 10 minutes. It's much slower than would be experienced at, say, a gas station, so be patient!

When the power at home fails due to any reason, what do you do? Sure, you have flashlights, don't you? (My recommendation is to have several CC Trek lights from ccrane.com strategically placed in easy reach at home, at the office and in each car's glove box, but I digress.) I find it handy to have a device, a sterling performer like this Energizer All-In-One around for peace of mind. There is a 200-watt DC/AC power inverter inside and an external plug, just like in the wall, to power small devices or a light, at home, on a camping trip, at the beach or in the yard. I like the idea of using a 7- or 8-watt LED bulb to provide as much light as a conventional 60W incandescent bulb (See the previous chapter to check out the 7W GeoBulb from ccrane.com as the best, most efficient of this new technology). One of these new bulbs can be operated during several dark hours per day for several days by a fully charged All-In-One; ideal for camping as well as emergencies. Add to this a flex-neck LED light to illuminate the area near the All-In-One for easy and safe hook-ups. When viewing the product online at the link above, notice the flex-neck lamp across the top of the handle, sitting in its own recessed track, and the switch just for the light at the base of the light, just above the number "2" that is clearly visible in the online picture.

There are TWO 12V DC outlets for your power accessories, too. Just like in your car, these outlets can be used to charge anything you would charge in the car, from phones to media players and more. A separate power switch energizes this section of the device. I like that! See it in the online photo in the POWER area.

When it is time to jump start a vehicle's battery, nothing I've used does it better or more safely. What is necessary to jump-start a vehicle? First, power. Inside this device is a sealed, lead acid, 18 Amp Hour battery, which is plenty for the task at hand *in most cases*. The safe way to do the job is to first clamp to the negative, "-" terminal on the battery OR clamp to ground; a nearby chassis bolt, for example. Next, a clamp goes to the battery's positive terminal. On this Energizer-branded product are a couple of safeguards that impress me.

NOTHING happens until the user turns the BIG knob to the Jump position. Notice it has a "3" next to it as you view the product online. This is to indicate that it is the *third step*. The first step is noted in the same area. See the number "1" there? It is a push button that tells the user if the battery inside is charged and ready to go.

Step "2" is placing the clamps, which I already touched on. What also needs to be mentioned is another intelligent design element. Note the rear of the unit. The actual product is slightly different, with a better-designed centrally located wrap-around device and the tips for the compressor hose neatly housed in the center of the wrap-around device.

Now, also observe the clamps on either end of the unit. Each clamp nests on these ends, clamped to a piece of plastic for this purpose in place at the top of each end. In this way, each clamp is stowed surely and safely, and ready for action, with each cable wrapped around the designated, color-coded guides, one for red (positive) and one for black (negative). It would be unlikely for a user to make a mistake.

And since nothing happens until the switch is turned to the "Jump" position, this means there is no danger, either, no unintentional power flowing. Another safety measure is in place. Should the user attempt to reverse the clamps, to place them red to negative and black to positive, there is an audio alarm that will sound to get the user's attention so this can be remedied. The "+" and "-" guides at each end, appropriately above their respective Positive and Negative clamps also tells the user what to do. Here is where the flex-neck LED comes in handy, too.

If there is not enough light, switching on this lamp while placing the entire device at rest under the hood and immediately next to the battery can show the user exactly what is needed.

Once the vehicle is started, it is easy to first switch OFF the jump starter, rendering the cables without power flowing through them and, therefore, safe. Removing the clamps is easy as it is easy to place them back as appropriate in their respective nesting places. Intelligent design resulted in clamps on THIS product that are tight, but not too difficult for use by *Mrs. Gadget*. Other products tested had clamps that are way too stiff for many females and, I am sure, for many men, as well. This includes being too stiff for me to comfortably use.

Here is another smart idea on this and not others purporting to do this job. It is taken for granted that the cables are an integral part of this product, that they are permanently affixed. On other, seemingly similar products, the cables are maintained NOT connected and not attached. The disadvantage is that they are kept separate, in their own plastic, zippered bag and must be plugged in to the main "box" for every use. I can tell you from experience this is a bad idea and all the more difficult for weaker users. I think of myself of at least average strength and even I have had difficulty plugging in the large cable connector on those products requiring this step. *Mrs. Gadget* found it nearly impossible. And once connected, the cable end protrudes below the level of the bottom of the box, making it awkward to set the box down. They got it right, way right, on THIS product! Score again for the Energizer All-In-One!

Now, as for the air compressor, this, too, is best-in-class design. Not shown well in any pictures to which I can point readers is the recessed area on the rear where the protective braid-covered hose stows. Take my word on this. The hose nests well in its appropriate place and the various nozzles, needle fillers and such fit neatly in a nest, ready to go, in the space on the rear of the unit between the cables. In the online photo, that area appears to be blank, but it is not on the actual product.

Charging the unit is also the result of intelligent design. Notice what appears to be two prongs of the type one would find on the end of a plug? That's it! To charge this product, all that is needed is a standard extension cord plugged *onto* those two prongs. Intelligent circuitry inside allows the unit to be plugged in all the time without fear of overcharging. OR, users may wish to plug it in and observe the green ready light on the front as indicative of it being fully charged and then unplug it. Plugging it in once overnight every three months is sufficient to keep it topped off. Were I you, I'd just leave it plugged in. The amount of current drawn to keep it charged is, I believe, negligible. ALL the other units in this category are difficult to charge. They rely on one of those brick-like charging adapters that must be connected under a hatch that is difficult to unlatch. This arrangement is clean, neat and easy as can be. Simply use your own appropriate length extension cord. While I have not done this, I am sure the unit may also be charged, or partially charged from a vehicle's own 12 V port, through a sub-200W inverter. Such inverters plug into the power socket and provide 120 V AC power for small-drain devices, including a product such as this. Ideally, one would not choose this method to charge an All-In-One unless conventional AC power is not available for an extended period, while the need for the All-In-One is more likely.

I've used ours many, many times over the past six months and it has delighted me and other users without reservation. From experience, including enlisting the kind assistance of *Mrs. Gadget*, I can tell you even SHE has remarked on how much better and easier this product is to use than any of the others we have tested. We ran down one car's battery regularly in order to test this and the other products in the category. Testing is concluded and the victor revealed.

We've pumped up bicycle tires and car tires, BIG exercise balls and sports balls for the kids in the neighborhood. One factor stands out. Through all its use, we never depleted the All-In-One's battery. We were always able to jump-start five times before recharging. We never tried to jump-start so many times that the battery inside became depleted. We figured five jumps would be more than one might require before recharging. We filled TWO auto tires from nearly flat without the All-In-One pooping out. That was running nearly 25 minutes without letup. We powered a small 60-watt equivalent CFL requiring 13-watts of AC electricity for more than 12 hours without the battery inside needing to be recharged. We charged iPods and phones and laptops and generally put this product to the test, never seeing its battery depleted to the point that we noticed it at minimal power.

As an emergency tool as well as a convenience tool, I am well satisfied with the performance, as well as the ease-of-use and intelligent design.

I won't embarrass the other brands tested in this category by naming them, but I will review their shortcomings.

The others found with similar battery capacity and with the other main features came with a separate bag in which are jumper cables, AC charger, DC charger and tips for the air compressor's hose end. To use the jumper cables, the user is required to plug the giant cable connector on the cables into the mating end on the battery box. There are four problems with this scenario. First is the requirement to keep track of that separate bag with the cables (and other accessories). Next, it is difficult, even for me, to firmly, fully attach the cables on the connector. Then, there is the fact that said connector is downward facing and that, when connected to the cables, the whole mess sticks downward, preventing the box from being able to be placed on a flat surface. What were the designers thinking!! As the connection becomes LIVE as soon as connected, extreme caution must be exercised so as to prevent the positive and negative cable clamps from inadvertently touching the wrong place. Oh, and in addition to the aforementioned issues, the cable clamps are exceptionally tight, difficult even for me, let alone *Mrs. Gadget*, to squeeze and clamp in place. It is as if designers never have had to use their product in order to see just how poorly it is designed and how difficult it is to use.

The best advice I can offer is to get this one, and only this one, clearly the best, most efficient and safest design among all competitors, and, as I found, also the best value. I found it best priced at Amazon.com for only $120, including FREE delivery.

Though the battery inside is a standard type and readily available, the product is not made to be user serviceable as far as I can determine. That is not to say that some enterprising person cannot do it. It is not something I would recommend attempting and when this one dies, I will simply purchase a replacement.

If earlier tests prove consistent, you can expect this product to perform admirably for at least five years, perhaps more. That has been my experience with others that use essentially the same battery type and capacity.

CHAPTER 15

WINDSHIELD

IF YOU DRIVE IN THE RAIN,
YOU NEED AQUAPEL ON THE WINDSHIELD!

Read this quickly, then buy and apply **Aquapel** (**aquapel.com**) to your vehicle's front glass to see clearly, more safely, longer than with any other rain-repelling glass treatment.

Ever the skeptic, I had nothing better to do one day many months ago than to look for something better than **Rain-X** (**rainx.com**). *But, why?*

Many years ago I met a man because of some technology I was investigating at the time. During our meeting, the man told me he was also involved in a product and its technology that was used by the aircraft and airline industry that protects their cockpit glass to make rain bead up and roll off like water off a duck's back. I said, you mean, like Rain-X? He laughed a good, deep belly laugh and said that was what he always hears. No, said he, Rain-X is "crap." Really, he said just that.

I listened intently as he continued. Rain-X is a great tool with great marketing. They know their product does its job only for a short time and they have conditioned their users to this. Rain-X requires reapplication with regularity, perhaps every two to four weeks in order to experience effectiveness. His chemical, on the other hand, bonds to the glass, won't wash off, is impervious to wipers and lasts for SIX months or more. I asked for some of this miracle juice!

Again, that was many years ago and I lost contact with the man, but I always remembered this product, having used it and coveted the small, unmarked bottle he sent. I nursed the contents and used it sparingly over the years until it was gone. Then, Rain-X was all I could find, so that is what I used. And Rain-X is so attractively priced, at way under $10 for a generous quantity. What's not to like?

For those of you currently using Rain-X, I am speaking to the converted. Inconvenience aside, it is somewhat effective at making driving in the rain much safer than without Rain-X on the windshield.

For those of you inexperienced with such things, imagine driving through the worst of the worst driving rainstorm, when your wipers cannot keep up with the volume of water pelting your windshield as you drive. Now, imagine the water beading up and being blown off the windshield as you drive at speeds over about 40mph. Imagine if you turned off your wipers in the heaviest rain and at this or higher speed, all would be fine, that you would see ahead clearly, without the distortion of sheeting water on the glass in front of you.

Imagine that the beads of water simply blow off your windshield and you can see clearly through the rain. Imagine no more, because it is real. It is the truth and driving in the rain is all the safer for this technology. The question becomes, "Why NOT use such a product?"

I will NEVER again drive without protecting my vision through the windshield so that I can see as clearly as possible even in the worst rainy weather. I make sure my family is protected, as well, and that all our cars are similarly prepared for the worst.

Unfortunately, Rain-X is drudgery personified. It's a chore to clean the windshield to its best, most spotless capability and then to reapply Rain-X. This must be done in dry weather, as well. It is inconvenient and time consuming. One must be careful to NOT get the chemical on paint, to carefully wipe it into the glass, then buff with a dry, soft cloth.

I'd rather do other things. To me, it is a tinker's dream to apply Rain-X. I don't want to tinker with this stuff any longer. Life holds other interests for me.

And that is why, when I saw a listing for Aquapel in my search for something better than Rain-X, I had to check it out. And so I did! No, I could not get more of my miracle juice, but Aquapel seemed to me to be similar in its claims.

The company behind it, PPG, was kind enough to send along several samples. First, I cleaned my windshield with a pot and pan scraping sponge and soap. I wanted it to be as clean and free of anything and everything as possible. That was in April of 2010. I knew I would not encounter rain here in SoCal any time soon, but I had other things in mind.

Knowing of the properties of chemicals such as this, and owing to my previous experience with super rain repellants, I wanted to test Aquapel in the face of bug debris one encounters while driving from home base through the desert to Las Vegas in hot weather. Those bugs come out and splat, they perish when hitting the glass at more than 80mph! Without Aquapel, cleaning the glass is not so easy. With Aquapel applied, the bugs and bug guts clean off much more easily. Mission accomplished.

And so I waited patiently, month after month, for rain to appear. In fits and starts, there had been sporadic rainfall during the following couple of months. Near the end of December, 2010 it began raining quite heavily. I was home for a couple of days nearing the end of my annual 15-day national TV tour during which I travel to other cities to show some of my top recommendations for holiday gift giving. I was so excited to be at home when it was raining!

Aquapel was *still* protecting my windshield! Now, I cannot say it will last as long for you on your vehicle's glass, but I can categorically state that it is NO contest with Rain-X. The two are chemically so different! I went out for a proof-of-concept drive. I drove to the freeway in the heavy rain and entered the nearest on ramp. At about 65mph and in the heavy downpour, I simply switched off the wipers. Amazing, still! I knew it would work this way before turning off the wipers. I could see how easily the water was squeegeed away by the wipers. I could see the water repelled above the wiper area. This stuff is remarkable!

Windshield wiper use, glass cleaner and car washes do not affect Aquapel. It is not silicone-based as is Rain-X nor is it removed through car washing. It chemically bonds to the glass, just as did the mystery chemical from many years ago.

Why would anyone want to use Rain-X and have to reapply it so often when Aquapel is so clearly superior? Aquapel is not six times more expensive, though it clearly lasts up to and maybe more than six times as long. And what about all that time you will save applying Aquapel once during the time others have to reapply Rain-X six or more times? After Aquapel, Rain-X is like water!

Now, **18 months later, that first application is still effective on my car**, though I will soon reapply in time for the upcoming so-called rainy season, such as it is here in SoCal. This is quite remarkable!

In the strongest way, I encourage, plead and beg of you to do as I and as I suggest – for safety's sake, get Aquapel and apply it to at least the outside of the front glass on your car and on the cars of those you love and care about. I can promise you all will see a remarkable, transformative difference, never again wanting to drive, especially in the rain, without this protection. Oh, and be sure the wiper blades are of good quality and in good shape.

If you wish, you may also apply Aquapel to the outside of the side and rear glass, and this is up to you, your degree of interest and budget.

Aquapel is not at all expensive, even if not considering its longevity of service per each application. Even if your situation and circumstance suggests reapplication on a quarterly basis, you are effort ahead of using Rain-X, which after only a week or so becomes much less effective than many, many weeks and now months after applying Aquapel.

You may be wondering why Aquapel is not a household name and number one in market share and popularity in its field. Let this be another lesson, please, that it is not always the best product that is most popular. I can think of many similar analogies in which the best at what it does in not the market leader. This has only to do with marketing. Rain-X is better marketed. That is the ONLY reason it sells so much better than the absolutely superior Aquapel. I wish I could be with each and every one of you in the first rain after your application of Aquapel. You will have an Aha! moment, I promise you.

Shop online for Aquapel to find the best price. The *retail price* is about $10 for one application, though you'll find it online for much less. When at the company Website, note the upper right area in which a Zip code may be entered in order to search for a local dealer. Give that a try, also.

Each applicator has a soft, felt-like pad. When the "wings" of the applicator are squeezed together, a glass ampule inside breaks, releasing the chemical onto the pad. Now, just spread it on the pre-cleaned glass PROMPTLY, and wipe the glass with a dry paper towel, not allowing Aquapel to dry on the glass. Done. (Instructions are at **aquapel.com/how-Aquapel-works.php**.)

I've just checked online for prices of Aquapel and found it at amazon.com and on eBay.com. Depending upon the quantity, the best price is at one or the other. **Priced from as low as two for $10 on eBay, with free shipping, to $25 for six on eBay, also with free shipping. Larger quantities are available from both sources.**

Here's to safer driving, courtesy of the remarkable Aquapel!

STEVE KRUSCHEN

CHAPTER 16

TRAVEL TECH ESSENTIALS

MY TRAVEL TOOLS, TRIED AND TRUE

Whenever, wherever consumers travel, there is more to think about than packing the right clothes! Of course, consumers will be thinking of travel costs. With higher gas prices than anyone would like, wacky and sometimes outrageous airline charges in addition to ticket prices, the craziness of TSA airport screening and a perpetual weak dollar overseas, travel costs continue to rise. I've got tips to help make the most of your travel dollars.

From luggage and luggage locks to electrical needs and mobile phones, even flashlights, I've got you covered with advice for the cost-conscious and just plain smart traveler.

Luggage – Yes, **Costco** sells *everything*, including luggage, but you may wish to think more carefully about your luggage needs with an eye toward longevity and not just price. Do look at **Costco's Kirkland Signature** brand for comparison, as well as the other brands sold by Costco. Similarly check at **Sam's Club** and **Wal-Mart**. If the traveler is not likely going on more than one luggage-toting trip per year, and for only a couple of weeks at a time with only one or two occasions during that time to encounter luggage handlers, the inexpensive brands may suffice. On the other hand, if lots of travel is the likely scenario, then it may be a good idea to spend the money for heavier-duty luggage that can stand up to whatever impediments may come its way. As a seasoned traveler, I've learned the hard way the value of sturdy luggage.

Look at the warranty – Of course, lifetime, no-questions-asked is best and many of the luggage makers are now providing this coverage. That's the good news. On the other hand, and I hate to be the *buzz kill*, if your luggage regularly requires free repairs, it's a hassle you can do without. Also, some of these warranties require that the consumer pay to ship the damaged bag to the repair facility, and those costs add up quickly. But it's the reliability I crave along with the great warranty. No one can prevent accidental damage caused by faulty handling on the part of your transportation carrier, but having quality luggage built to take it and built to last will avoid many of those problems. With longevity in mind, I look for semi-hard-sided, ballistic nylon luggage with a soft top, wheeled of course, and with pull-up handles. That's just *my* preference. Many travelers are opting for four-wheeled ("spinners") soft (or hard cases) and these, too, are good choices, though I don't see any advantage in hard-sided spinner-style luggage, which I will soon explain.

Where to buy - My choice is to shop for *the good stuff* (not the really expensive bags), which can last through many, many years of rigorous travel and require fewest repairs. I have always found the best brands at the best prices right here in Southern California at **Savinar Luggage (savinarluggage.com)** in **Canoga Park, CA**. If you are in the Los Angeles area, visit them at **6931 Topanga Canyon Boulevard**, off the Ventura (101) Freeway, two blocks north of Vanowen Street on the west side of the street. Call them at **818-703-1313 and please let them know *Mr. Gadget*® sent you.** Either take their advice over the phone (as I do) and let them ship directly to you or visit them in person. Let them suggest the brands and models that meet the traveler's needs, whether it is **Delsey (delseyusa.com)**, **TravelPro (travelpro.com)**, **Tumi (tumi.com)**, **Swiss Army (swissarmy.com)**, or **Briggs & Riley (briggs-riley.com)**, as well as many other recommended brands they carry. If you think the best prices, even on sale items, will be found visiting a luggage store at the mall or the luggage department in a department store, forget about it. **The prices and selection at any other retailer I've found cannot match what I have found at Savinar,** in business since 1916. Even if Savinar is not near to you, first shop and price compare in your area. Then, call Savinar and they will likely beat the best price you found, including shipping. I've found they even beat **ebags.com** luggage prices on the bags I want to buy.

On the topic of spinner-style luggage, the four-wheeled bags that stand upright and move along with little effort on those wheels, there are two styles.

One is a more traditional design of sturdy ballistic material over hard sides, with a soft-top and structurally hard bottom, with a pull-up handle. The other is called hard-sided. These are created without the fabric, with hard plastic material making up both the top and bottom surface of the bag. The bag is like a clamshell, with the zipper equally dividing the halves. As the hinge opens from top to bottom, as the bag stands, it cannot be packed nor opened in a conventional manner.

Imagine placing your conventional-styled bag on a luggage cart or on a bench at the foot of the hotel bed. You'd place it with the handle on the right or left, as dictated by the zipper opening's orientation. The top would swing up from the front, then become vertical at the back, perhaps leaning against the wall behind the luggage cart.

Contrast this with the high-styled clamshell spinner, which splits in two vertically. It cannot fit on the luggage cart. There is significant added width to accommodate. In addition, each half of the clamshell is just that, an equal half, each of which is to be packed separately, more carefully and with less height than when packing the deeper single-well of traditional design. It is not convenient and may be impossible to pack a box, for example that is thicker than one of the clamshell halves. In a traditional design, taller box heights are more easily packed, without the need for concern of one half or the other. As I said, I just don't get it. Do you? And yet, as I travel, I see many consumers with these clamshell designs. It is something the traveler should consider when shopping for new luggage.

While on the topic of luggage and what to look for in your next purchase, consider the expandable bag, the one with a zipper or some other means by which to temporarily increase the thickness and, therefore the capacity of the bag. Be sure the wheels of a spinner or the bottom bumpers is not a spinner are under the expansion and not on the unexpanded portion. If not on the expansion, when the bag is in expanded mode, it will not be balanced to easily stand upright as before. A brand I have come to rely upon for outstanding value, thoughtful and careful design, solid, long-term quality and their famous no-questions-asked warranty is **Briggs & Riley**. Their bags are designed with features appreciated by seasoned travelers. **It is the brand I have chosen for my most recent luggage set.**

Luggage locks – HIGHLY recommended (I never travel without using them), but the *right* ones are essential. Whatever your choice, make certain they are TSA-approved. This will assure that your locks *should* not be cut off by the TSA (in the US only, by the way) during inspection. There are numerous TSA-approved locks from which to choose. **The only ones I use and recommend are SearchAlert-type combination travel padlocks by Sesamee (sesameepadlocks.com – then click on Products, then Sesamee Travel Security).**

These locks permit users to set any three-digit code. TSA (Transportation Security Administration) personnel have keys that allow them to open these locks for the purpose of luggage content inspection. A Security Window next to the SearchAlert name shows red to indicate entry has been made using one of these override devices. Easily re-set to green, ONLY by the consumer, travelers will know instantly at the destination if a lock has been opened so the contents of the attached suitcase can be checked before leaving the airport. **ONLY SearchAlert locks by Sesamee have this critical feature.** SearchAlert locks retail for about $10-12, so let Google find them using **searchalert locks** as the search term. Savinar also carries them. Put locks on each luggage compartment where there is a provision for one on the zipper pull, so you may need as many as three or more on each piece of luggage, plus a spare or two.

TIP: Don't forget to put locks on your lockable carry-on bags. You never know when you might be told there is no room in the cabin for a bag. The lock in place means you are ready to hand over a carry-on bag for a gate check. Without that lock, baggage handlers can easily help themselves to the contents (another lesson I learned the hard way!).

Portable digital luggage scale – It is bad enough that the airlines offer less service with higher prices, but most charge a fortune for luggage. And they watch more carefully than ever the weight of that luggage. Do not EVER get stuck with overweight luggage. Weigh before you go! On nearly every trip, I see some poor sap at the airline counter frantically removing packed items from a suitcase in order to come in under the weight limit. This slows up everything for everyone at airline check-in as well as for everyone waiting in line, and can be an embarrassment for the frazzled traveler at the counter as well! Get an inexpensive portable digital luggage scale to avoid surprises and to be more efficient all along a travel itinerary. My choice is **Balanzza Mini Luggage Scale (balanzza.com)**, under $25 online. Weighing about three ounces, it's easy to use and goes along inside a bag so it can be used during the trip, too. Just strap the scale to your luggage, lift using the scale, wait for the beep, set luggage down and read the weight on the digital display.

Before leaving the general topic of luggage, there is another way to avoid having airlines lose or damage your luggage. Consider shipping bags ahead to their destination using common carriers such as FedEx and UPS. This recommendation comes from noted travel expert and former colleague, **Peter Greenberg (petergreenberg.com)**. He practices what he preaches, too, and has done this for many, many years. He *flies* through airport security without luggage. He rarely has to fight for overhead space, either. His luggage is at the destination hotel, awaiting his arrival. Peter says the cost may not be much different, in many cases, from what the airlines charge for checked baggage. It's worth investigating!

Flashlights – I urge everyone to carry at least the two flashlights that I carry. If you've ever been caught in a strange place in the dark (or if you NEVER want to be caught in the dark), you will appreciate this recommendation and solution immeasurably.

Everyone needs one of these inexpensive pocket lights that are indispensable to travelers, domestically or internationally. They're great for everyday use, too. Keep one on your keychain and give one for the keychain of everyone you love.

By far, the best, brightest little light on the planet is the **Photon Micro-Light II**. This little lifesaver lights the way in a powerful and indestructible end-of-your-thumb-size package. Traveling or not, I believe in and rely on this product. I have been told the same product is also used by the US military, Secret Service and even aboard NASA's Space Shuttle and the International Space Station. You can get the same super-bright light with long-lasting, inexpensive, replaceable Lithium batteries! The amount of light that comes from the Photon Micro-Light II is quite amazing. **Accept no substitutes! There is simply nothing like it or better than the Photon Micro-Light II**. You'll see other small lights, seemingly competitors, for less and be tempted to buy, but don't do it. Get genuine Photon Lights ONLY. Most users need the Micro-Light II with the standard black body and white LED (as opposed to any of the other color LEDs available). Battery life is up to about 12 hours, though by then it will be considerably dimmer. Still, you'll get many hours of useful light before needing to replace the pair of inexpensive batteries. **Available for less than the $12 suggested retail price at Savinar and elsewhere**. Shop online, too. **Manufacturer's Site: laughingrabbitinc.com. The manufacturer's retail Site is at this link, shortened for your convenience: bit.ly/vB8oav**. Buying from the manufacturer's retail site assures you of getting the most up-to-date and brightest revisions and freshest inventory. **I like this light more than all** others because it is unlikely that it will be accidentally activated in a pocket or purse, then remain on and drain the battery, rendering it useless when needed most. Yet, there is still a mechanical On/Off switch for those occasions you want it to stay on. ALL others lack this dual capability or are of demonstrably inferior quality.

An alternative or additional light is the same company's **X-Light Micro**, a squeeze/click on and off design, with additional modes via micro circuitry inside. Read about it at the manufacturer's Site above. Notice that it has a suggested retail price of only $10. It is simple and reliable, but it lacks the hard on/off switch of my favorite model, the **Photon Micro-Light II**. This means it *can* accidentally be activated, that is, clicked *on*, and stay on, inadvertently depleting the battery. Still, it is a formidable performer.

The second flashlight I rely upon and recommend to travelers (and everyone) is the **$30 CC Trek LED Flashlight from my friends at the C. Crane Company** (ccrane.com). Once at their Website, click the link to LED Lights, then LED Flashlights to find this model. This US-made powerhouse has proven its worth many times for me, both while traveling and at home during power outages. **There is nothing better for the money, and nothing as good for less.** Four super-bright LEDs throw an amazing amount of light.

The **CC Trek** is great for taking along on dog walks, hikes or exploring antiquity in the dark. It's also exactly what you'd want to have with you to light the area in the dark for any reason, and especially for those unplanned times when you need light and lots of it for an extended period of time. **With fresh batteries (which are already installed), full brightness is maintained for at least 40 hours.** No other product can match this. If needed, it could be left on all night while you sleep in what otherwise would be total darkness and it would still be at its full brightness in the morning. You can easily see why this is the light of choice for travelers, including this writer. In addition, it is indestructible and waterproof to 160 feet. What's not to love? CC Trek is also perfect for the glove box and for many other uses at home. Three inexpensive and readily available AA alkaline batteries power this LED flashlight.

For those who want even more flashlight brightness and high tech capabilities, the industry has responded to the flashlight-aholic. Why even more power? From experience, higher brightness flashlights allow daytime brightness where it is needed and otherwise quite dark. I've used my favorite super-bright flashlights on hikes, to help others while camping, exploring cavernous buildings and caves and, dare I say, just for fun! **My favorites in this extreme category are by Chinese maker, Fenix, (fenixlightus.com).** I love their flashlights that use disposable AA alkaline batteries. Why? Because these batteries are inexpensive and widely available. Some Fenix high-performance flashlights, and those from other makers, use more expensive Lithium CR123 and similar disposable cells that are not the same size as AA and AAA. I think the best performance and *value* comes from flashlights using disposable, standard size AA and AAA alkaline batteries. In preparing for and anticipating certain travel needs, I sometimes replace the standard AA alkaline batteries with **Energizer AA Ultimate Lithium batteries** that offer greater runtime between the need for replacements. Yes, they cost more, but they last so much longer so as to add a convenience factor that may be worth the added cost for extreme users. It's great to have the flexibility of power sources these products provide.

The Fenix models I like and recommend for travel include the single AA-powered **LD10 R4, about $50**, and their two AA-powered **LD20 R4, about $60**. Both are multifunction and blindingly bright. Be sure to read about the multifunction characteristics! Shop online for the best price. **Read more about flashlights in the chapter dedicated to them!**

Powering your stuff – It sure would be nice if all the plugs and voltages in the world were the same, but, alas, they are not. Having what's needed for international and domestic travel will avoid the heartache of dead batteries and blown up products fed too much voltage during overseas travel. Convenience, voltage and the correct plugs combine to make your electronic equipment happy!

First and of greatest importance, look at the chargers for all your take-along products. You will need to see all the writing to determine the input voltage range. If it says "100 – 240 volts" as I suspect most will, you are all set and need only plug adapters and NOT *voltage converters* or *transformers* for your trip. For now, let's assume you are good to go. If you only see 110 volts listed as input voltage, and if you must take the device along on international travel, you will absolutely need to purchase and use a little, yet heavy, voltage adapter that steps down the 240 volts found overseas to the 110 volts of your product. More on this later.

Convenience dictates accessibility to a hotel room wall outlet that is where you can use it. For example, some hotel rooms are simply not equipped with conveniently placed wall plugs nor are there enough of them. Travelers need to charge mobile phones, camera batteries, mp3 players, electric shavers, electric rechargeable toothbrushes such as **Sonicare**, and computers, and maybe more. Note: Not all **Sonicare** toothbrush chargers are designed for international use. You may need to buy a Euro charger and may need to attach a plug adapter depending upon the country visited. When it comes to toothbrushes, it's best to just take a good old regular toothbrush unless you are certain that your electric toothbrush will operate properly either on a full charge without recharging for the duration of your international travels or that the charger is designed for international use with a plug adapter as described below.

With potentially so many things to plug in I have a solution that has worked for me everywhere, domestically and abroad. **Start with a common, inexpensive 15-foot indoor extension cord** with a standard two-prong grounded plug on one end and three outlets on the other. For this example, we will visualize the female side at the right. This is the end that things are plugged *into*, with one outlet on one side of its "block" and two on the other. **The 15-foot length will assure that the traveler will be able to access the outlet wherever it may be in the hotel room, yet use the other end, the business end with all those plugs, in the most convenient location in the room.**

I bought mine at the local **Do-It Center**, but you'll find what you need at most other hardware stores – **Ace, Lowe's Home Improvement, The Home Depot**, and others. Expect to pay about $6 for this type of 15-foot extension cord. **I chose a two-prong instead of a grounded three-wire cord because this one is less bulky in consideration of less bulk and weight to pack.**

Next, you'll need what I call a **triple tap**, also from **Do-It Center**, and also found at **The Home Depot** and **Ace Hardware** stores, among others. It has a three-way plug and three, three-way outlets *around its perimeter*. The outlet configuration allows acceptance of both three-prong plugs, such as what many computers use, as well as standard two-conductor grounded plugs such as the "bricks" used with cameras, mobile phones and iPods. By having them placed around the outside, one per side, the individual blocks plugged in will not interfere with each other. In-line adapters featuring three outlets side-by-side do not provide enough space between outlets for the big blocks described, so get and use only the triple tap adapters. This adapter costs about $5.

A **grounded adapter** is the final puzzle piece. It will be used connected to the single outlet side of the extension cord's three outlets, converting the three prongs of the triple tap to the two on the adapter.

Then, the two connected above are themselves plugged into that single side of the extension cord's three-outlet block. Do this and your kit is ready for use – 15-foot, two-prong extension cord with three outlets; three-way triple tap plug with outlets around the perimeter; three-prong-in-to-two-prong-out adapter. If you only need three standard two-prong outlets, you will not need these last two pieces for now, but do travel with all three pieces, separated until they are needed, and then separated again to pack them in your suitcase, carry on or computer bag. Keep them all with you to be prepared for any required combination – three standard outlets or as many as a combo of five if all the pieces are connected.

Use any required plug adapter (on the two-prong plug end) necessary for the country visited and then plug the assembly in the power outlet in the wall. In this way, there is just one plug in the wall outlet, with minimal weight and bulk. Many hotel room outlets are loose-fitting from wear, so you'll appreciate NOT having a bunch of things plugged directly into the wall. Plug the adapter stack in the extension cord. Next, plug in up to five items as would be done at home in the US, so long as their voltage requirements range from 100-240 volts. Four adapters that come in an inexpensive kit detailed below cover virtually any possibility in Europe.

I keep a reusable cable wrap with hook and loop closure (you may call it Velcro, but that is a brand name for hook and loop closures) to keep it all neat and tidy. They are indispensible for jobs like this. Search online for **reusable Velcro cable strap.** Keep all cords neat, untwisted and wound in an orderly fashion and they should last for many years. Don't let them become kinked or closed in doors or in any situation in which their integrity may be compromised. This is for safety's sake as well as to maintain general reliability. You do want it all to work, don't you?

An alternative to the flexible and durable wrap, and a good one at that, is the original Cable Clamp (cableclamp.com). These are simply the best! Buy a set for use on extension cords, garden hoses, rope and so many other bundles. They even have new little ones called **CableClic (go to cableclamp.com and click on the CableClic link)** for your earbud wires and other cords on small personal electronics and adapters.

International voltage - International travel to countries with plugs and voltage different than ours is not difficult to manage, but it requires care and planning in advance. Just attach a single plug adapter to the plug end of the extension cord, and plug it into the outlet. My extension cord "kit" provides up to five outlets as described above for up to FIVE devices that do NOT need voltage change. You only need to find one plug in the wall in whatever country you visit, and, of course, you need only one plug adapter for any country to plug in up to five products. There's less to pack with this system and less complexity. So simple!

In my example, the adapters on the left are **PLUG ADAPTERS**, not voltage adapters. Be sure to buy and take along plug adapters for all the countries to be visited. **Check with the resources below for your needs.**

Don't worry about overloading the circuit or the extension cord. All of your small electronic products as described above do not draw much current. In other words, the load is very light even when including charging a computer. **DO NOT** plug in an appliance such as a hair dryer or flat iron to this same extension cord along with other products.

International travelers are best advised to leave their hair dryer and flat iron at home. To operate overseas on that *other* voltage, a small, heavy adapter is required. Forget it. Why take extra weight and add to the complexity when it is not needed? Purchase a hair dryer and flat iron, if needed, which have a switch with a 125-volt setting *and* a 240-volt setting. They're designed with travel in mind! Find them at **Magellan's** (**magellans.com**) in their catalog online or in their Santa Monica or Santa Barbara, CA, stores. Do NOT get the models with their own adapter *plugs* as you will not need them. Travelers may use the plug adapter from the extension cord setup described for the short time it would be needed to dry or iron the hair. Remember to disconnect anything else plugged in while using a hair dryer or flat iron. Then, when finished with the appliance, remove the plug adapter and put it back on the extension cord. **TIP: switch to the 240-volt setting prior to packing to avoid mistakes that most assuredly would blow out the product the instant it is plugged in and turned on with the wrong voltage setting.**

If you find that you *do* need to step down the visited country's 240 volts to power your 110-volt only product, you must get a converter, the *correct* converter for the job. You'll have to look at the info plate on your device, and then look online at **Magellan's Voltage Converter page** or call them at **1-800-962-4943** for assistance. You may also try their **Electrical Connection**

Wizard page at **this shortened link: bit.ly/v7qdN3** and see if you can help yourself.

One final admonition – DO NOT PLUG IN ANY PRODUCT WHILE TRAVELING OVERSEAS WITHOUT BEING CERTAIN OF ITS VOLTAGE REQUIREMENTS **BEFORE LEAVING HOME**. IF A MISTAKE IS MADE, EVEN IF ONLY ONCE, THE ELECTRONIC PRODUCT WILL BE TOAST IN A NANO-SECOND.

Phones, phones, phones – This topic is the most confusing and confounding for most consumers who travel abroad. I wish I could report that it is easy, but, sadly, it is not. Do your homework! The result will be a much less costly alternative than a surprise bill from a home carrier for many hundreds of dollars, or worse!

A mobile phone used today in the US *might* operate overseas. Then again, it *might not*. Unfortunately, there is no simple solution that can apply to all consumers. It is more than likely that your **US-based Sprint mobile phone** will not, cannot work overseas, and if it can, you will be locked into using their expensive service even when in foreign lands. It is likely that your existing **Verizon phone** cannot operate overseas, unless it is among few dual-band models also equipped with **technology called GSM**, the only standard in place just about everywhere in the world except for the US. **AT&T Wireless** and **T-Mobile** customers' phones are similar, though they do not use exactly the same GSM technology. These handsets and they *may* have the right stuff to enable overseas service. **Ask your carrier, but beware, they will only know if that handset can operate, perhaps with voice only, with the existing SIM card inside.** I know, this may be confusing, but let's move along regardless.

The good news is that the **AT&T** and **T-Mobile** handsets of US consumers may work overseas. The bad news is that, if your handset will work, **the rates using your existing carrier's SIM card for making calls back home and for receiving calls from home, in addition to making and receiving calls to and from callers within the country visited can be excruciatingly expensive.** Check with your carrier for all the information on pricing plans and the availability of services – voice, text, Internet and email, multimedia messaging and any other services of interest. Oh, and be prepared for a shock when the cost is revealed as well as a not so easy to understand array of variables.

There *is* another, better way to go for the **occasional international traveler**. It is possible to sign up with services designed for travelers at great rates as compared with using US-based carriers' plans. These services allow the use of existing **GSM** phones capable of being used in visited countries or the services can supply a purchased SIM card or a purchased handset with SIM card designed for the use of occasional business and vacation travelers.

In order to use an existing US-based **GSM** handset, the phone must be "unlocked," that is, the **GSM carrier** (**AT&T** or **T-Mobile**, for example) must enable that it will recognize an installed foreign SIM card. Your **AT&T**- or **T-Mobile**-branded handset is sold to customers with this feature blocked. "Good customers" of **AT&T Wireless** and **T-Mobile** may get their domestic phones "unlocked" in order to use them while abroad with a foreign country SIM card IF the handset (phone) is of the correct technology on the inside. Just because it works here does not mean it will work over there. Call or visit your carrier's retail location to ask about foreign compatibility and for details on how they can enter the appropriate code to electronically unlock the handset. I've heard tales that retail location personnel will not always be helpful to walk-in customers and they may not admit to their customers that the handset may be unlocked. If in doubt, call your carrier at "611" from the mobile handset and speak with a customer service representative! It is likely that the carrier will provide instructions via email detailing exactly how to unlock your **GSM** phone. I've done this with all our **AT&T**-branded mobile phones so we are always prepared for international travel.

This will NOT work with AT&T iPhones. AT&T will NOT unlock iPhones. There is a way to get what you want, according to the Apple representative with whom I just spoke. Purchasers of a new iPhone 4S, who get it for use on AT&T can travel to any GSM ONLY country (Europe and countries on that side of the world), remove the native AT&T SIM and pop in a local one. Whatever services come with that SIM should work on that iPhone. Were I you, I would still check and confirm the above well in advance of travel.

Alternatively, consumers may purchase unlocked **GSM** handsets either directly from some manufacturers or from non-carrier owned retailers selling phones and services. These handsets are sold at a premium because the carrier is not subsidizing their use. In other words, there is no guarantee that your purchase of an unlocked handset will result in its successful use through a particular carrier, as you would most certainly use an **AT&T**-branded handset with your AT&T Wireless account.

ASK about and be certain of compatibility well in advance of leaving on your trip!

My personal preference is for unlocked handsets from **Sony Ericsson** (**sonyericsson.com**) that are not tied to a carrier and are, in my view, cooler than many phones sold by the carriers. Go to **sonyericsson.com** and click on **Products** to see all the **SonyEricsson** phones. Tri- and Quad-band phones should work in most foreign **GSM** markets, but always confirm with the manufacturer to get what you need, should you also want to use the phone in the US on either **T-Mobile** or **AT&T**, or on both. Again, you will see there are many handsets available directly there that are not available through US carriers.

Check the Website (**sonyericsson.com**) to see styles and learn about features. Then, call them toll free at **1-866-766-9374** to get advice on which models will perform the song and dance of your choice. I've also found their personnel on the other end of information and tech calls to be exemplary.

Locked and unlocked **Motorola** phones are available directly from **Motorola** through **store.motorola.com**.

Other handset makers, to my knowledge, do not sell directly to consumers. That is, to buy their brands purchases must be made from the carrier.

In addition to the option of buying a carrier-locked GSM phone and having it unlocked by the carrier or buying an unlocked carrier-independent phone and using a domestic plan when overseas, following are those *other* available solutions.

Call in Europe (**callineurope.com**) offers mobile phone solutions for European travelers that make the most sense to me and are at the best rates I've seen. **I have used and have been pleased with service and plans from Call in Europe. This is also the solution of choice for students studying abroad in Europe and in other countries of the region.**

The company advises the following:

First, check to see if your cell phone will operate overseas by asking the carrier – AT&T or T-Mobile.

If your current provider is Verizon, Sprint, Alltel, Telus or any other CDMA carriers, your phone will not work there (with few exceptions that your carrier can detail). You have the choice to purchase a *travel kit* that includes a European cell phone and phone line that is ready to work – with a starting price of $58 at callineurope.com. This is certainly your most economical option. You will receive a GSM phone that you can keep and reuse for future trips overseas. There is no need to return it and you'll benefit from the lowest rates available in the marketplace with per-minute calling rates well below those charged with a rental product.

If your current provider is AT&T, T-Mobile, Rogers or another GSM carrier your phone might work in Europe; however, you will need to confirm if your device is either a tri-band or a quad band phone as most European networks operate on 900/1800 Mhz. If this is the case, then you will need to contact your US or Canadian carrier and ask for the unlocking code of your device. This code will be provided to you for free in the US so long as you have been a customer with that carrier for at least three months. Once your phone has been unlocked, then you will simply need to get a Call in Europe SIM card providing you with the ability to make and receive calls at the lowest rates available in the marketplace.

To choose the SIM that will give you the most benefits, you will have the choice between a local prepaid card (offered by the carriers of the country that you are planning to visit), a European prepaid SIM card (cellularabroad.com, telestial.com) or a European postpaid SIM card (brightroam.com, callineurope.com). Most of them may be conveniently purchased prior to departure. Among those previously listed, the most convenient option with the best quality of service and the lowest rates would be the European postpaid solution furnished by Call in Europe.

There will be no need to buy calling credits in advance that may expire if you don't use them or to refill your card in the middle of a conversation. Call in Europe offers the lowest rates available in the marketplace conveniently billed in US dollars to protect against foreign currency exchange rate fluctuations. With prices that are 50% less than a local prepaid SIM card, 70% less than any US or Canadian carrier with no contract, no commitment nor deposit requirement, the savings offered by Call in Europe cannot be easily overlooked.

Call in Europe is partnering with the leading European carriers and offers a direct dialing solution not to be confused with a call back system (which is often not as reliable and tends to be more costly and less attractive). Incoming calls are free in France or in the UK, and $0.39/min elsewhere in Europe. Outgoing calls are $0.39/min to call locally, as well as call to another European country or back to the US/Canada. Text messaging is free to receive and between $0.19 and $0.39 to send depending on your location in the world. To compare with a US/Canadian carrier incoming/outgoing calls start at $0.99/min, whereas with a local prepaid solution, incoming calls tend to be free; however, local calls and calls to the US are between $0.80 - $1.20/min. International prepaid SIM card solutions offer free incoming calls; whereas, outgoing calls are often provided at rates from $0.50 - $0.99/min. (Rates and other charges cited may have changed, so contact them via the above link for the most up-to-date info from Call in Europe.)

The Call in Europe solution is also **ONE of TWO** European SIM solutions that offers both data and voice access for Blackberry and other Smartphones. **Roamsimple.com** is the other.

Thank you Call in Europe! Smart ideas, reasonable rates and solutions for any traveler.

Here is additional information from Call in Europe and further food for thought.

• Remember that is it cheaper to receive calls in a foreign country when calls are from someone in the US. This suggests that, whenever possible, call back that traveler on their foreign phone or plan.

• US carriers may filter text messages from Europe to the US, so a test is in order to be sure it works. I am advised that there should not be a problem texting to **AT&T Wireless and other US carriers**, but be forewarned.

• Text messaging to the US and to others that may also be in France, for example, costs 19 cents each. Texting from Europe outside of France to others in Europe outside of France may be at a cost of 39 cents per message.

• Incoming text messages to the foreign travelers in Europe using **Call in Europe** are free, though they are NOT free for us to send from here to there. Ask your carrier for the rates that apply to you.

• Calling while in Europe to others in Europe has varying costs. For example, calling within France, mobile to mobile, is 69 cents per minute.

Calling to others in other European countries while outside of France may cost 99 cents per minute. There is no uniform rate.

• Calls from Europe back to the US from France, for example, cost 39 cents per minute and 99 cents per minute to the US from other European countries.

• Sending a simple email from a handset in Europe to the US may be less expensive than sending text messages to the US!

The biggest take-away from **Call in Europe's** advice is that there *is* a cost-effective way to keep in contact with the US and with others traveling in Europe. In addition, we should stay clear of the inconvenient and costly pre-paid callback scenarios – the traveler calls home to the US, for example, and the US party must call back through a cumbersome series of numbers that ultimately rings the phone in Europe. Who wants *that* hassle! I think we just want to be able to call and be called with the best price and greatest convenience, and it seems that **Call in Europe** has the best, most cost-effective solutions available to the short-term European traveler.

If your travel plans call for visiting international destinations outside of Europe . . . The best alternative to using home-based GSM mobile phones in other locales, such as Australia, Asia, Africa, the Middle East, India, and Russia is **brightroam.com**.

What about Verizon or Sprint customers? The technology of their wireless phones is called CDMA, completely different than and incompatible with GSM. According to Sprint and Verizon reps, 4G CDMA smartphones should work all over the world, but only using Verizon or Sprint services or roaming in the other countries. Translation? A big phone bill! See above for Call in Europe solutions for purchase of a *travel kit* or buy an unlocked GSM phone for this temporary travel, then purchase a SIM card through Call in Europe.

iPhone 4S is different, designed as a world phone, with technology inside for CDMA as well as GSM carriers. I quizzed an Apple sales representative, asking, "What if an iPhone 4S, which initiates service through Sprint or Verizon, is considered for travel?" According to the representative, if an iPhone 4S begins service as a Verizon or Sprint device, it CAN have a GSM SIM installed in another country and it should work. Once back home, it will revert to being its original CDMA phone on its original network.

Again, check with your carrier! You might want to check with your carrier before buying your next phone to learn about foreign compatibility **NOT** using your US carrier. Even if you only occasionally travel out of the country and want to use your own phone with a foreign SIM inside, I want you to avoid the shock of a huge phone bill upon returning home to the US.

Taking a digital camera? If your smartphone is not sufficient, take along a dedicated digital camera, the most modern of which can take not only stunning still photos, but also high definition videos. Not all offer continuous autofocus with zoom in video mode, so look for one with this capability.

Be sure to take a large enough memory card or more than one. Not only will you want to get loads of *still* photos, most of today's digital cameras are also capable of taking high definition videos with sound. The camera's manual will have more information or contact the manufacturer.

When it comes to brands, you're in luck! Many manufacturers offer excellent products that won't break the bank: **Simple point-and-shoot, high on the megapixel count, small in size, great in low light, large LCD display, at least 3X optical zoom, excellent videos *with zoom*, good to excellent battery life, inexpensive high capacity camera memory, $200 range street price.** Low light performance is critical for consumers, in my view. Sure there's a flash, but it does not carry far on any pocket camera. Great low light performance makes it possible to capture the moments of life indoors and out, in the wide lighting circumstances of life, without flash much of the time, and without the graininess older cameras exhibit without this feature. Fortunately, many modern digital cameras are getting better, even excellent results in very little light. Good, inexpensive memory is important. It is the combination of features above that are in the cameras I think are best in the $200 range.

Look for cameras from **Canon, Nikon** and **Panasonic**, as well as **Olympus,** as your best bets. Oh, and I should not forget about **Sony** pocket-size digitals in their CyberShot line, though I have not had much experience with them of late. It's difficult to find a bad one among any of these brands.

My personal preference based on the most experience is for **Canon pocket-size models**, many of which are simply excellent performers in all aspects, including the ability to take 720p or 1080i high def video. Amazing!

I want to recommend an additional important feature on just about any digital camera that is also particularly useful for travelers – a wide-angle lens. This features allows the shooter to be closer to the object yet still take it all in from left to right and up and down. The ideal specification is 24mm wide. Some are only 28mm wide, but try to find one you like with the 24mm feature. Imagine, for example, standing at the corner of a large dining room table and being able to snap a photo of everyone at the table in one shot. Or think about traveling to the great spots in Europe, in the Holy Land or in exotic Asia, in front of ancient sites. Now, imagine being to be quite close, yet to be able to capture the full scope, left to right and top to bottom of some of the greatest buildings and other objects known to mankind, without having to stand back so far that all detail is lost. THIS is why I recommend purchasing a camera equipped with a 24mm wide-angle lens. And of course, it should have at least a 3X zoom lens at the very least. A better number would be at least 5x, but the combo or high zoom with 24mm wide angle is not common.

Today's cameras may use high capacity memory called SDHC. Newest cameras may also accept SDXC, the newest, highest capacity SD cards, perfect for those who want to maximize a combo of massive numbers of digital stills and lots of high def video, perhaps hours, all on one card!

TIP: Be prepared for travel with enough memory to allow you to take tons of photos *and* videos. Why not, it's FREE to take them on reusable memory cards and you don't have to print any of them. With only one 8GB card, a 10-megapixel camera can take more than 1,000 pictures at best quality, plus videos. With a mix of photos and short videos, determine if one card will meet your needs.

Higher megapixel cameras will need one or more cards of greater capacity. The "sweet spot" in SDHC camera memory today seems to be 16GB cards. I just looked and found my favorite Kingston brand in a 16GB Class 4 card for only $13, with free shipping. Class 4 is a speed rating, the right speed for most point-and-shoot cameras as well as most digital SLRs, including video cameras. Check specifications for YOUR camera, though, then shop! I know that careful online shopping yields significant savings on camera memory, so don't wait until there is no choice but to pay more at a local shop or while on a trip.

Take multiple cards and use a permanent marker to indicate 1, 2, 3 or A, B, C on each so you can easily identify which is which. Once a card is full, avoid heartache by **LOCKING** the card until it is time to remove the photos. **All of them have this sliding switch.**

I buy, use and recommend camera memory cards made by Kingston Technology (kingston.com/flash/photo.asp). (I also find **Kingston** to be among the most reasonably priced name brands for reliable computer memory on the market.) Of course, there are other quality camera memory brands from which to choose, including **SanDisk**, **Lexar** and **PNY**, to name a few. The higher the megapixel count on the camera, the greater is the need for a larger card – a 10-megapixel camera's photo will be about 3 megabytes (MB) in size, for example. Videos need lots of megabytes, but they are so worth the effort. And high def videos eat up even more memory.

Go ahead and take along plenty of memory for long trips so you can shoot and shoot and shoot some more, then bring them all back home to off load onto your computer. Remember, seeing **high def video and preserving on optical media** are two different sides of the same coin. Most of us do not have **Blu-ray** recorders to create high def DVDs, but save the files anyway as native high def, then create standard DVDs as needed using your computer's software. Whenever it is that it becomes cost effective to create **Blu-ray** DVDs on our computers, those files will remain and be able to be re-made in native highest quality onto **Blu-ray** DVDs! In the meantime, streaming these videos on your high def TV is becoming quite popular. It may come to pass that home-grown high definition videos remain on a hard drive and are played back ONLY from a hard drive.

So long as the high def videos remain on the memory cards, the *camera* can play back directly on a connected **high def TV**, using the proper cable. Consult your camera's manual for more on this, but most use a mini version of **HDMI** at the camera end and a standard size **HDMI** on the TV side. In addition, many modern computers feature some method of outputting the native high def files for display on an outside high quality monitor. Many of today's high def TVs feature this computer interface, so, again, consult your manual!

TIP: When the day's or event's shooting is done, it's time to get the photos and videos off the card and transferred onto the computer, if possible. If you're on vacation without the computer, this should be done soon after returning home. Once the transfer is made (using a cable from the camera to the computer or using a separate memory card reader) and you are certain all went well and that you can access all the photos and videos, it is time to dump them from the memory

card to free up that space for more fun. With the memory card back in the camera, access the camera's menu and navigate to the setup menu to locate the Format Memory Card function. Consult your camera's manual if you do not know how to do this. NOTE: Format the card each and every time photos are transferred to your computer and/or you have shared them with others who may want the shots for themselves on their computer. Formatting the card makes the card behave as if fresh and new, preventing errors that can ruin your day! Do this without fail and keep your memory cards happy and healthy and reliable for many years and untold thousands of photos in the future.

Camera batteries? – If your digital camera uses "AA" size batteries, there are two good options.

One is to get the new rechargeable so-called "hybrid" batteries that come pre-charged and hold a charge for up to 365 days. No other rechargeables are recommended. Unfortunately, using any rechargeable batteries while traveling is troublesome. The user must remember to manage perhaps daily charging, taking a charger and multiple sets of batteries. If this is your choice, I have tested and can recommend the newest Lenmar (pre-charged and ready to use right out of the package) AA and AAA rechargeable batteries (lenmar.com, then hold cursor over BatteryCentral, and at bottom of grey window that opens, enter the word rechargeable in the search box and click search. Select either AA or AAA as required, then – be certain to look at descriptions once the link opens since you want the "pre-charged" rechargeable batteries). Duracell Pre-charged Rechargeable AA batteries (duracell.com, then enter rechargeable pre-charged in search box on top right) are another good choice, but generally cost more. Be sure to get the ones that say Pre-charged on the label! Find them at all the popular retailers, or shop online for better prices. *These batteries can be charged hundreds of times* with any battery charger designed to charge standard Nickel Metal Hydride (NiMH) batteries. Other brands with similar technology include Sanyo Eneloop (us.sanyo.com/Battery-Products) and Rayovac (rayovac.com, then click on Rechargeables).

For cameras using AA-size batteries, a more manageable solution for travelers, though more expensive, is Energizer disposable AA Ultimate Lithium batteries (energizer.com). These are power-packed and worry-free because there is no charging to manage. If the camera takes a pair of AA batteries and normally would get, say, 200 shots on a set of charged NiMH batteries, it would not be unreasonable to expect up to 500 shots on a pair of *disposable* **Energizer Lithium batteries**. An eight-pack retails for as little as about $17 at local stores. Shop online, too, at retailers including **thomasdistributing.com.**   Shop with my friends there for not only Energizer Ultimate Lithiums but also for the "hybrid" AA cells.

If your camera is supplied with a rechargeable Lithium battery – Most small and popular point-and-shoot digital cameras come with slim, rechargeable Lithium batteries. The good news is that the chargers are most assuredly set for worldwide voltage. In addition, replacement **rechargeable Lithium batteries for *most* cameras are available from Lenmar, and they carry a two-year warranty (lenmar.com,** then click on **BatteryCentral** to find your camera and its Lenmar replacement Lithium battery).   Copy the Lenmar product number that corresponds to your camera's battery, then **search online** for the word "Lenmar" AND the Lenmar part number you find for that camera's replacement battery.

You will be amazed at the prices found. **I often suggest that consumers do what I do; purchase a pair of two-year warranted Lenmar replacements. With prices as low as $25 for two Lenmar batteries delivered to your door, how can you go wrong?** Shipping is generally the same price for one or two batteries.

This way, the consumer can use the original charged battery and then easily pack two more that will hold their charge for many weeks when not in use. Be prepared and avoid being caught without a charged battery for your camera, especially when on a trip far from home.

TIP: If you have multiple extra batteries, use a permanent marker to write a 1, 2, 3 or A, B, C on each for easy identification. Try to rotate their use so each battery gets exercised, that is, used and recharged, so all will be ready to go when needed.

Now that you are equipped with a small, point-and-shoot digital camera, extra **batteries and memory, how about a small camera case that can hold everything but the battery charger?** Here is what I do. I gather all the pieces together and head over to Best Buy where I find Lowepro camera cases. I buy the smallest Lowepro case that fits the camera, at least one extra rechargeable Lithium battery (or set in the case of AA) and at least one extra memory card in its case.

Recharge USB-capable devices in a car, anywhere in the world. Whether traveling in the US or around the world, any mobile device capable of being charged using USB can be charged using a vehicle 12V power adapter. This is ONE worldwide voltage standard, thankfully. There are many good quality, dual-outlet USB vehicle chargers available, so why not be equipped in the car at home and in the event you can make use of the plug while on far-away travel in other vehicles? Charge a phone, iPod, iPad, and other USB-rechargeable products. One example is from **Lenmar**, their **model AIDCU2**, routinely found online and at retail stores for less than $15. This one also is equipped with an easily replaceable 2A car fuse.

CHAPTER 17

HYDRATION

HYDRATION BOTTLES -
THE BEST TO QUENCH YOUR THIRST

The trend toward personal hydration bottles – "water bottles" – started several years ago. I can remember the days when I was thirsty, I visited a water fountain or found a way to get a cup or glass of water. Don't you? It was not the big deal it is today.

Today, it is commonplace to see people toting their water bottles everywhere, in and out of doors, at home and at the work place and while exercising. What happened? I cannot begin to guess, but clever marketers seized another opportunity to make us beholden to something new – water bottles. Save the environment with reusable bottles!

As long as we're hooked, I wanted to have a look at standouts in this field. I found three that are different and I think *better* than plain old reusable water, I mean *hydration* bottles, as well as bottles designed for frigid cold water and piping hot beverages. The common thread is their quality and leak proof design or exemplary ability to maintain hot or cold liquids over extended periods of time.

First up, Intak from Thermos (thermos.com). These inexpensive and simple bottles are best for room temp water and feature a positive locking and leak-proof design in 18- and 26-ounce styles. Ice inside will make these bottles sweat.

Intak's seal on their first design is a simple, positive, click-lock design. A press of a big button pops the top to reveal the solid spout on this first style. There is a metal clip that may be employed to hold the top over the bottom for an even more positive locking seal. Another style uses an inset sealed top that pops to reveal a built-in flip-up straw. Take your pick!

Cost is as low as $5 up to about $10. The material is BPA-free, impact-resistant and dishwasher durable **Eastman Tritan copolyester** (plastic).

Tritan is my favorite material for this use! It absorbs no tastes or odors and is virtually infinitely reusable.

You know the drill – search online for best pricing!

Next are the many and varied styles from **Contigo**, also using Tritan (**gocontigo.com**). Contigo's leak-proof **AUTOSEAL** products are made for active lifestyles and one-handed operation in capacities ranging from 14-ounces to 21-ounces. The leak-proof AUTOSEAL design comes from a proprietary spring-loaded seal that is released and opened with one hand's grip, making them ideal for active lifestyles and drivers. Grip and squeeze to open, and release to seal. Some users with hand mobility issues have reported their lack of grip strength makes it difficult to press hard enough to open the seal.

Also of special note is their new double wall insulated **Martinique** 18-ounce bottle. What a smart idea! The double wall design is made from the same Tritan material I like. With the double wall design, this one *resists* sweating when filled with ice-cold water and maintains the cold for up to two hours. It's not sweat-free in all circumstances, but it *is* a noticeable improvement over single wall designs. Typically selling for $13 online, the cost premium seems a worthwhile up sell.

AUTOSEAL *vacuum* insulated 16-ounce mugs of stainless steel may be used to maintain steaming hot beverages for about four hours and ice-cold beverages for up to 12 hours. Vacuum insulated designs are completely sweat proof and stay cool to the touch, whether filled with hot or cold liquids. These are leak-proof designs starting as low as about $6!

For users favoring a straw-type design, **Contigo's AUTOSPOUT 24-ounce Addison Water Bottle** is just what you've been waiting for! A press-to-release button pops up the silicone waterspout on this one-handed bottle. The spout it neatly tucked away and protected when retracted. Though not part of the AUTOSEAL line, there have been NO leaks in my testing, so long as the spout is not released. Made from the same durable Tritan material as the other Contigo bottles I like and recommend, this one-handed bottle is sure to be a welcome addition for the specified user. I found it online for less than $15.

Occasionally I have spotted triple-packs of **Contigo AUTOSEAL** Tritan bottles for about $20 in Costco stores.

The best of the best, leak-proof, sweat-proof and cool-to-the-touch bottles no matter what's inside is relative newcomer Hydro Flask (hydroflask.com). **So good and so simple are these that they require little explanation.** Now just three years old as a company, theirs are premium stainless steel double-wall vacuum insulated bottles that **maintain HOT beverages for at least 12 hours and ICE cold beverages, *with ice*, for as much as 24 hours.** Hydro Flask bottles are nothing short of amazing, and they are made so well! Find them in 12-, 18-, 21-, 24-, 40-, and 64-ounce models, priced from $20 to $60. All are unlike anything I have seen or used, and **clearly the best available. Each is available with a standard loop cap or with a pull-up sports cap and comes with a limited lifetime warranty, as do all the other brands and types featured here.**

See all recommendations through the links provided, shop online and choose what's best for your needs. They also beg to be gifted.

INDEX

T

tablet · 7, 31, 42, 43
television · 1, 188
theft protection software · 81
THX · 12, 13
triple tap · 163
tuner · 2, 11

V

vacuum · 21, 105, 106, 107, 179, 180
VoIP · 18, 52, 92, 97, 98, 99
voltage adapter · 162

W

webcam · 33
wide angle lens · 172
WiDi · 7
Wi-Fi · 6, 17, 18, 28, 30, 32, 35, 42
Windows · 3, 7, 8, 11, 31, 33, 34, 37,
38, 39, 40, 41, 43, 49, 51, 56, 57, 58,
59, 60, 62, 65, 66, 67, 69, 70, 74, 75,
76, 77, 79, 81, 85, 91, 94, 95

STEVE KRUSCHEN

The One and Only Mr. Gadget®

ABOUT THE AUTHOR

Steve Kruschen, the one and only Mr. Gadget®, is the guy who tries it before you buy it! For more than three decades, Steve has been testing, demonstrating and reporting on consumer electronics, gadgets and new technology. He checks out products in a real-world, hands-on environment -- just the way the average consumer would. As Mr. Gadget, Steve helps consumers make intelligent buying decisions by providing radio, TV and Internet audiences with his trusted evaluations of the latest and greatest. Thanks to his numerous radio and TV appearances, including *Fox & Friends* (FOX), *The Early Show* (CBS), *Early Today* (NBC), *Power Lunch* (CNBC), plus appearances on CNN, The Weather Channel and dozens of news shows in major cities, consumers rely upon the expert advice of Mr. Gadget. When he is not appearing on television, Steve is speaking to corporate and trade association audiences on how to improve productivity with personal technology. Steve lives in Southern California with his wife, *The Long Suffering Mrs. Gadget* (and their three children — when they visit). Batteries not included.

www.ingramcontent.com/pod-product-compliance
Lightning Source LLC
Chambersburg PA
CBHW072229270326
41930CB00010B/2050